Matthias Keller

Formation of Intracardiac Electrograms under Physiological and Pathological Conditions

Vol. 21
Karlsruhe Transactions on Biomedical Engineering

Editor:
Karlsruhe Institute of Technology (KIT)
Institute of Biomedical Engineering

Eine Übersicht aller bisher in dieser Schriftenreihe
erschienenen Bände finden Sie am Ende des Buchs.

Formation of Intracardiac Electrograms under Physiological and Pathological Conditions

by
Matthias Keller

Dissertation, Karlsruher Institut für Technologie (KIT)
Fakultät für Elektrotechnik und Informationstechnik, 2014
Referenten: Prof. Dr. rer. nat. Olaf Dössel,
ao. Univ.-Prof. Dipl.-Ing. Dr. techn. Ernst Hofer

Impressum

Scientific
Publishing

Karlsruher Institut für Technologie (KIT)
KIT Scientific Publishing
Straße am Forum 2
D-76131 Karlsruhe

KIT Scientific Publishing is a registered trademark of Karlsruhe
Institute of Technology. Reprint using the book cover is not allowed.

www.ksp.kit.edu

Print on Demand 2014

ISSN 1864-5933
ISBN 978-3-7315-0228-9
DOI   10.5445/KSP/1000041272

Formation of Intracardiac Electrograms
under Physiological and Pathological Conditions

Zur Erlangung des akademischen Grades eines

DOKTOR-INGENIEURS

von der Fakultät für

Elektrotechnik und Informationstechnik

des Karlsruher Instituts für Technologie (KIT)

genehmigte

DISSERTATION

von

Dipl.-Ing. Matthias Walter Keller

geb. in Reutlingen

Tag der mündlichen Prüfung: 15. Mai 2014
Referent: Prof. Dr. rer. nat. Olaf Dössel
Korreferent: ao.Univ.-Prof. Dipl.-Ing. Dr.techn. Ernst Hofer

Danksagung

Diese Arbeit entstand im Rahmen meiner Tätigkeit als wissenschaftlicher Mitarbeiter am Institut für Biomedizinische Technik (IBT) des Karlsruher Instituts für Technologie zwischen August 2010 und Mai 2014. Hiermit möchte ich mich ganz herzlich bei allen bedanken, die mich während dieser Zeit unterstützt und damit zum Gelingen dieser Arbeit beigetragen haben. Im Folgenden möchte ich einige unter ihnen besonders hervorheben.

Zuallererst danke ich ganz herzlich Herrn Prof. Dr. rer. nat. Olaf Dössel für seine sehr gute Betreuung sowie konstruktive fachliche und moralische Unterstützung bei der Durchführung meines Forschungsprojekts und schließlich für die Übernahme des Hauptreferats.

Des Weiteren danke ich Herrn ao.Univ.-Prof. Dipl.-Ing. Dr.techn. Ernst Hofer für seine umfassende Unterstützung und Beratung bei Fragen zum experimentellen Teil meiner Arbeit sowie für die Übernahme des Korreferats.

Bei Herrn Dr.-Ing. Gunnar Seemann bedanke ich mich für die kompetente Betreuung des Simulationsteils meiner Arbeit.

Meinen Kollegen am IBT danke ich für eine stets angenehme und produktive Arbeitsatmosphäre. Für eine tolle Zusammenarbeit in verschiedenen Projekten danke ich besonders Dr.-Ing. Mathias Wilhelms, Tobias Oesterlein und Gustavo Lenis.

Für eine sehr produktive und spannende Zusammenarbeit im Bereich der klinischen Signalanalyse danke ich Herrn Dr. med. Armin Luik und seinem Team vom Städtischen Klinikum Karlsruhe.

During my stay at the Cardiovascular Research and Training Institut I was able to gain a lot of experience in the field of experimental studies of cardiac electrophysiology. Therefore, I would like to thank Prof. Dr.-Ing. Frank Sachse, Chao Huang and the whole group for a great and exciting time at the University of Utah.

Ich danke Herrn Dr.scient.med. Robert Arnold für die erfolgreiche Kooperation und die tollen gemeinsamen Experimente. In diesem Zusamenhang spreche ich auch meinem Labor-Team Ramona Modery und Dirk Falkenberg meine tiefste Dankbarkeit für viele gemeinsame Versuchsstunden aus.

Zahlreiche studentische Arbeiten haben zum Erfolg meiner Forschung beigetragen. Wegen ihres großes Engagements möchte ich hierbei Franziska Schäuble und Lür Tischer hervorheben. Ganz besonders danke ich Steffen Schuler für seine tolle Bachelorarbeit und seine langjährige sehr erfolgreiche Hiwi-Tätigkeit.

Für die Unterstützung bei der Entwicklung von Komponenten des Versuchsaufbaus und deren Fertigung bedanke ich mich ganz herzlich bei der Werkstatt des Instituts, sowie bei Herrn Manfred Schroll.

Ich danke meiner Familie und meinen Freunden, dass Sie in dieser spannenden aber auch sehr anstrengenden Zeit immer für mich da waren und es mir ermöglichten, auf andere Gedanken zu kommen und Kraft zu schöpfen.

In dieser gesamten Zeit hat mich meine zukünftige Frau Alexandra durch Höhen und Tiefen begleitet, mich stets unterstützt und mir Kraft gegeben. Dafür danke ich ihr von ganzem Herzen.

Die Forschung in dieser Arbeit wurde von der Deutschen Forschungsgesellschaft (DFG) im Rahmen des Projekts DO637/12-1 gefördert.

Contents

List of Abbreviations

AF	atrial fibrillation
AP	action potential
APD	action potential duration
AS	active segments
CathAblation4mm	ablation catheter with a 4 mm tip electrode
CathAblation8mm	ablation catheter with an 8 mm tip electrode
CathRing1mm	ringelectrode catheter with 1 mm electrode spacing
CathRing2mm	ringelectrode catheter with 2 mm electrode spacing
CFAE	complex fractionated atrial electrograms
Cross-Catheter	ablation catheter with one centered ablation electrode surrounded by four measurement electrodes
CS	coronary sinus and coronary sinus catheter
CV	conduction velocity
dV/dt_{max}	maximum absolute deflection between (p_{pos}) and (p_{neg})
ECG	electrocardiogram
EGM	electrogram
Φ_e	extracellular potential
Φ_i	intracellular potential
IEGM	intracardiac electrogram
UEGM	unipolar electrogram
BEGM	bipolar electrogram
KHS	Krebs-Henseleit solution
LAT	local activation time
Micro-Catheter	ablation catheter with three integrated microelectrodes

NLEO	non-linear energy operator
p_{neg}	maximum negative peak
p_{pos}	maximum positive peak
para-dp	propagation direction from the distal to the proximal catheter electrode
para-pd	propagation direction from the proximal to the distal catheter electrode
PSD	power spectral density
GUI	graphical user interface
RFA	radiofrequency ablation
ROI	region of interest
V_{e}	extracellular potential
V_{m}	transmembrane voltage
V_{pp}	peak-to-peak amplitude
VSD	voltage sensitive dye
WCT	Wilson's central terminal

1

Introduction

Cardiac arrhythmias can cause life threatening conditions and are one of the major causes of death in the western world. The number of patients suffering from these cardiac diseases increases with progressed aging of the population [1–3]. Fibrillatory excitation patterns in the atria and ventricles prevent the heart from homogenous contraction and therefore obstruct the pumping mechanism of the heart. In case of the ventricles, this directly affects blood supply of the circulatory system, which can lead to death within minutes. The most hazardous side effects of atrial fibrillation are embolic events which can lead to strokes. The mechanism behind these events are blood coagulations, which form in the atria in zones of reduced flow due to inhomogenous contraction. Therapies for cardiac arrhythmias are anti-arrhythmic drugs and implanted defibrillators in the case of ventricular fibrillation. A more and more common minimally invasive method to cure patients suffering from arrhythmic cardiac diseases is the ablation therapy. In this procedure, cardiac excitation patterns are analyzed measuring intracardiac electrograms with multiple catheter electrodes inside the heart [4]. The aim is to identify triggers, patterns and substrate for fibrillation and isolate or deactivate these regions by precise generation of lesions. Lesions are supposed to block cardiac excitation between diseased and healthy areas and thereby restore and stabilize normal function of the heart.

The key to a successful ablation therapy is the correct identification of the arrhythmogenic substrate based on intracardiac electrogram analysis. New developments in catheter design and manufacturing as well as improved clinical data acquisition systems have paved the way for a detailed analysis of multiple electrode signals. Ongoing research has shown, that not only the annotation of activation times, but also the shape of intracardiac electrograms can reveal information about the underlying cardiac tissue [5, 6]. Therefore, a clear understanding how pathological excitation patterns are reflected in the shape of intracardiac electrograms is essential. Many theories exist, how specific electrogram patterns found in clinical electrograms can be translated to cardiac excitation patterns. However, the understanding of the underlying mechanisms is still poor and controversial.

Computer simulations of cardiac electrophysiology offer the unique opportunity to study how specific excitation patterns on pathological substrates are reflected in simulated intracardiac electrogram morphology. In order to use this technique to support clinical diagnosis, a thorough validation of the used approaches is necessary. For this purpose, datasets of clinical or in-vitro data are required, which include a good knowledge about the recording conditions, which can then be reproduced in simulations for improved parametrization. Validated computer simulations can help to understand unknown excitation patterns seen in clinical data and therefore support and improve the therapy of cardiac arrhythmias. Furthermore, new catheter designs can be developed and optimized in a fast and cost effective way.

1.1 Focus of the Thesis

This thesis is focussing on an improved understanding of intracardiac electrogram formation with respect to the characterization of cardiac tissue surrounding the measurement electrode. Therefore, the two main aspects of this thesis are:

- The design and validation of a detailed simulation environment for intracardiac electrograms

- The development of a setup for in-vitro measurements, in which electrograms can be acquired on vital cardiac preparations and at the same time, excitation patterns can be spatially characterized by recordings of a voltage sensitive fluorescent dye from the surface of the preparation

Pursuing these goals, three different fields of interest can be identified, which are covered in this thesis:

Experimental Studies

- Development of an experimental setup, which allows to perform optical mapping of cardiac excitation patterns of small tissue preparations using a voltage sensitive fluorescent dye
- Design of a tissue bath, in which vital tissue preparations can be studied using electrical and optical techniques under stable temperature conditions
- Design of a sensor for electrical measurements of extracellular signals, which resembles the geometry and materials of a clinical mapping catheter

Clinical Studies

- Acquisition and analysis of annotated clinical electrograms under physiological conditions as basis for validation of simulations
- Investigation of intracardiac electrograms during radiofrequency ablation sequences

Simulation Studies

- Establishment of a small scale simulation environment including detailed geometries of clinical catheter electrodes
- Validation of the simulation approach with clinical and experimental electrograms
- Investigation of electrogram changes due to physical and physiological parameters, as catheter orientation, conduction velocity and cardiac wall thickness
- Study of electrograms under pathological conditions including fibrotic patterns and radiofrequency ablation lesions

- Test of new catheter designs for improved radiofrequency lesion assessment

1.2 Structure of the Thesis

Part I introduces medical, modeling and technical fundamentals forming the basis of this thesis:

- **Chapter 2** gives an overview of the physiology of the human heart and the mechanisms of atrial fibrillation.
- **Chapter 3** reviews effects of radiofrequency ablation on electrophysiological parameters of cardiac tissue.
- **Chapter 4** summarizes the basics of intracardiac electrogram acquisition and analysis.
- **Chapter 5** introduces the basics of experimental studies of cardiac electrophysiology. The techniques of optical recordings using voltage sensitive fluorescent dyes and electrical measurements are explained. And the basic aspects of in-vitro studies of vital cardiac tissue preparations are given.
- **Chapter 6** outlines the basics of modeling studies of cardiac electrophysiology including cardiac cell models and the mathematical description of cardiac excitation propagation.

Part II presents the methods developed in this thesis:

- **Chapter 7** introduces the experimental setup for in-vitro studies of cardiac electrophysiology, including an optical recording system and the developed electrical measurement sensor, as well as peripheral components to keep the preparation vital. The bachelor theses of Franziska Schäuble and Hanno Gao and the diploma thesis of Lür Tischer contributed to the development of the presented experimental setup.
- **Chapter 8** presents the datasets of acquired clinical data as well as methods for automized electrogram analysis.
- **Chapter 9** explains the parameters of the developed simulation environment for simulations of intracardiac electrograms.

Part III presents the results of the performed studies:

- **Chapter 10** displays simulation results obtained from clinical catheter geometries under physiological conditions. Different influencing parameters on electrogram morphology, such as filter settings, catheter orientation and conduction velocity are investigated. Furthermore, simulated data is compared to clinical recordings for validation of the method. Parts of this chapter were published in the proceedings of the conference "IEEE EMBC 2013" [7] the proceedings of the "BMT 2013" [8] and the "Computing in Cardiology 2012" [9].

- **Chapter 11** describes results of in-vitro experiments and compares simulation data to measured electrograms.

- **Chapter 12** presents a simulation study on the influence of a measurement electrode on cardiac electrophysiology.

- **Chapter 13** covers a simulation study of the influence of fibrotic patterns on intracardiac electrogram morphology. Results are compared to a database of annotated clinical signals. The master thesis of Mohammad Soltan Abady formed the basis for this study, which was published in the proceedings of the conference "Computing in Cardiology 2013" [10].

- **Chapter 14** introduces the results of a developed modeling approach for acute catheter ablation lesions. Simulated and clinical electrograms during radiofrequency ablation lesion formation are studied. Furthermore, experimental results on conduction velocity development under hyperthermic conditions are presented. The bachelor thesis of Steffen Schuler contributed to this chapter. Parts of the shown results were also published in the "IEEE Transactions on Biomedical Engineering" [11].

- **Chapter 15** presents simulation results obtained from two potential future catheter designs for optimized radiofrequency ablation lesion assessment.

- **Chapter 16** summarizes the results and concludes the thesis.

Part I

Fundamentals

2

Physiological Basics

In the first part of this chapter, the basics of anatomy and electrophysiology of the heart are explained. Afterwards, atrial fibrillation is introduced as the most important cardiac arrhythmia. The therapy of atrial fibrillation is the most relevant application of the presented results in this work.

2.1 The Human Heart

The human heart is a four chamber organ, which pumps blood. Thereby, it forms the basis for blood based convective transport in the body (Fig. 2.1). It incorporates two separate pumping units, which are each formed by atrium and ventricle. Blood is pumped towards the lung (pulmonary artery) and the systemic circuit (aorta). Returning blood enters the heart by the venea cavea (superior and inferior) and the pulmonary veins [12].

2.1.1 Functional Physiology

Oxygen depleted blood coming from the systemic circuit is collected in the two venae cavae. These veins open into the right atrium, from where the blood is transferred into the right ventricle. By the pumping of the right ventricle the blood reaches the lungs. After the passage, oxygen rich blood flows into the left atrium and the left ventricle. The left side of the heart is responsible for the supply of the systemic circuit through the aorta. Furthermore, the heart itself is perfused by the coronary vessels, which emerge from the onset of the aorta. The direction of blood flow in the heart is ensured by valves

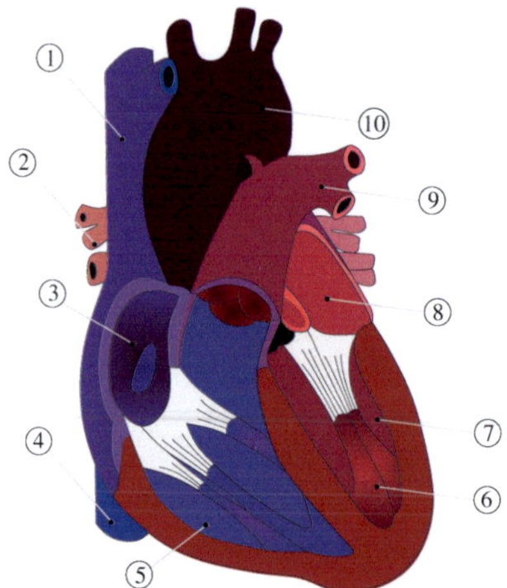

Figure 2.1. Anatomy of the human heart. (1) Vena cava superior, (2) pulmonary veins, (3), right atrium, (4) vena cava inferior, (5) right ventricle, (6) papillary muscle inside the left ventricle, (7) left ventricle, (8) left atrium, (9) pulmonary artery, (10) aorta

between atria and ventricles and at the outflow tracts towards the pulmonary artery and the aorta [13]. The mitral and tricuspid valve, separating atria from the ventricles are connected to the small free running papillary muscles by thin connective fibers [1]. Contraction of the heart is triggered by an electrical depolarization wave, which is passed on from cell to cell all over the myocardial surface. Under physiological conditions this depolarization is initiated by a trigger area in the right atrium, called sinus node. Starting from the sinus node, both atria are depolarized. The valve plane electrically uncouples the ventricles from the atria, except for one conductive bridge, the atrioventricular node. While passing the node, the propagation is delayed and high frequent activation of the ventricles is prevented. The depolarization is rapidly passed on over the conduction system of the ventricles, which is formed by the bundle of His, two bundle branches, and finally the Purkinje fibers. This mechanism ensures, that contraction of the ventricles starts at the apex, pushing the blood through the outflow valves[14].

Cardiac electrophysiology is based on two mechanisms. On the one hand cardiac cells are excitable, which stands for their ability to change their transmembrane voltage upon excitation by an electrical stimulus. On the other hand, cells are electrically connected to each other, forming a functional syncytium. [14].

Transmembrane voltage of cardiac cells (V_m) in rest is at about -90 mV. This voltage is established by an imbalance of ion concentrations between intra- and extracellular space. Resting ion concentrations are established and maintained by an active transport protein, the sodium-potassium exchanger (Fig. 2.2). It transports sodium out and potassium inside the cell. The transmembrane voltage is further established by potassium channels which are open at rest. These channels allow potassium to leak out of the cell, till an electrochemical equilibrium between transmembrane voltage and concentration gradient is reached.

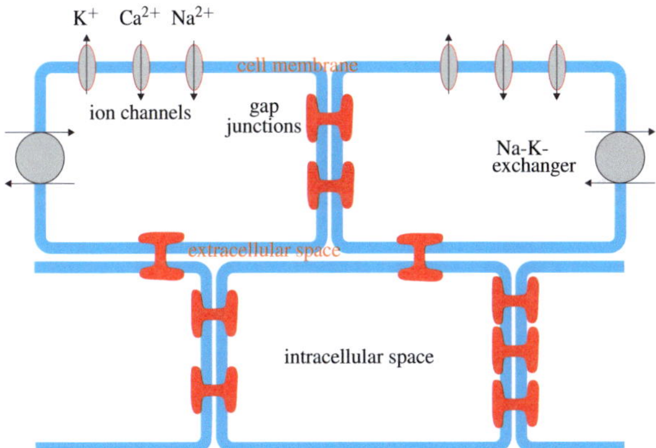

Figure 2.2. Basic components of myocardial tissue. Myocardial cells are outlined by a cell membrane (blue), which incorporates membrane proteins, such as ion channels, ion exchangers or gap junctions (red). Through the connection by gap junctions, the intracellular space forms a continuous compartment. The compartment outside the cells is formed by the extracellular space.

Cardiac cells are connected by small intercellular channels, which are called gap junctions. Thereby, a connection between the intracellular spaces of the myocardial cells is formed, which allows direct exchange of ions and small molecules [15]. Changes of intracellular ion concentrations or capacitive charging of the cell membrane leading to a more positive transmembrane voltage initiate an action potential (AP), (Fig. 2.3) [12, 14].

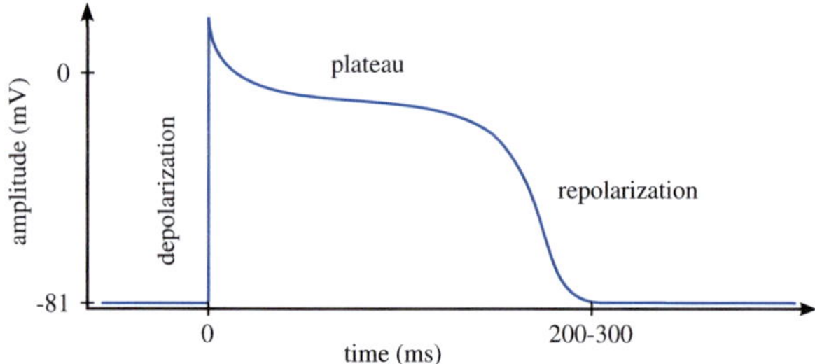

Figure 2.3. Schematic of a human cardiac action potential. Transmembrane voltage is depicted over time. The duration of APs varies throughout the heart. Atrial APs are shorter, than those in the ventricles.

The AP is initiated by transmembrane voltage gated sodium channels. Sodium influx leads to a rapid depolarization and also is the main source for excitation of neighboring cells [15]. Sodium channels close time gated [14]. During the depolarization phase, potassium channels open, which leads to a repolarization to a value of about $0\,\mathrm{mV}$ (Fig. 2.3). The second phase of the AP is called plateau phase, in which potassium outflux and calcium influx are balanced. After closing of the calcium channels, potassium based repolarization returns V_m to the baseline level.

The calcium influx serves two important physiological mechanisms. Firstly, increased calcium levels in the intracellular space activate binding sites for the contractile elements of the cell. Secondly, the prolonged AP duration makes the cell unexcitable (refractory), which helps to prevent self sustaining activation reentry patterns (fibrillation) in the heart [12, 13, 15].

2.1.2 Cardiac Wall Structure

The heart is positioned inside an elastic sac, the pericardium. Its interface towards the heart wall is formed by a thin fluid layer, providing smooth movement of the heart. The wall of the heart muscle can be subdivided into three layers. The outside of the heart wall is formed by the epicardium, which has a high content of connective tissue, fatty tissue and blood vessels. The myocardium is the thickest of the layers. Its main components are cardiac muscle cells, which are aligned along a dominating fiber direction. The endocardium entirely covers the inside of the cavities of the heart. Its inner most layer is formed by endothelium, followed by underlying connective tissue. The endocardial thickness of the ventricles is thinner than that of the atria [16]. For several areas of the left atrium the sum of epi- and endocardial thickness was reported to be 0.93 ± 0.63 mm compared to 2.91 ± 1.06 mm for the myocardium [17].

2.2 Atrial Fibrillation (AF)

Atrial Fibrillation is clinically described as a supraventricular tachyarrhythmia. Due to its chaotic excitation patterns, it prevents coordinated mechanical function of the atria. AF can be categorized as paroxysmal (self terminating), persistent (terminable by cardioversion) or permanent (not terminable). AF itself is not a life threatening condition. Nevertheless, the risk of stroke and other thromboembolisms is highly increased [18].

2.2.1 Potential Mechanisms Causing AF

AF can occur without detectable underlying mechanisms or along with an associated cardiac comorbidity [18]. In many cases AF was found in combination with ectopic foci in the atria. These are described as areas outside the sinus node, which initiate excitation waves. Frequently, these foci were found at the ostia of the pulmonary veins of the left atrium [19]. Furthermore, a substrate promoting reentry paths is a necessary condition for AF. One aspect in this context are increased levels of fibrosis found in patients with long standing AF [20–23]. Along with fibrosis also the interconnection of myocytes by gap junctions is altered [21], which can influence propagation patterns in the atria.

It has been reported, that AF itself leads to transformations in the atrium, which sustain this arrhythmia. This remodeling process has three targets which all promote reentry patterns of atrial excitation. Smaller action potential duration shortens the refractory period. Therefore, the so called wavelength is reduced. The wavelength is the length of an excitation path, which fulfills the condition, that the wavefront reaches excitable cells, after returning to the starting location. Furthermore, the atria are dilated and thirdly increased fibrosis levels provide the opportunity for labyrinth-like excitation paths [24].

2.2.2 Fibrosis

Fibrosis is described as collagenous inlays in between myocardial fiber strands. In the healthy heart, collagen supports the structure of the heart and ensures homogenous distribution of the contraction force [25]. In the pathological case, collagenous mass increases drastically. This increase is assumed to be caused by activated fibroblasts, which mediate the production of collagenous tissue [26]. Different patterns were found in human heart preparations. Compact fibrosis forms large connected areas. Small homogeneously distributed fibrotic inlays were described as diffuse fibrosis. Fibrotic elements are often aligned to myocardial fiber direction. If these elements only consist of fine strands, they were named interstitial fibrosis, compared to patchy fibrosis for increased element thickness [27]. Fibrosis was reported to affect velocity and patterns of conduction [27–29]. The arrhythmogenic potential of different fibrotic patterns was rated to be unequal. Compact fibrosis forms areas which are depleted of myocytes. Therefore, it is considered the most benign type. The only pro-arrhythmic mechanism related to compact fibrosis is the creation of long excitation paths encircling these areas. Interstitial and patchy patterns lead to an uncoupling of muscular strands, which creates zig-zag paths of excitation propagation. This effect is assumed to be a major cause of arrhythmias. Diffuse fibrosis was also classified as possible arrhythmogenic substrate due to its altering effects on propagation and cell-to-cell coupling within the myocardium [26].

2.2.3 Treatment of Atrial Fibrillation

In most cases AF is treated by pharmacological therapies. These include drugs to control the heart rate and anticoagulation therapy [18]. Among the non-pharmacological therapies, ablation therapy has been established as a method of choice during the past years [2, 5, 19, 30–33]. The aim of all ablation techniques is to cause myocardial cell injury in a defined and local way. This is done using different sources of energy, which are tissue cooling (cryo ablation), laser irradiation, ultrasound induced damage and radiofrequency currents. The most important among the listed techniques is the use of RF-currents, which produce local tissue heating at the location of a tip electrode [4, 34]. This technique and its effects on the myocardial cell properties are reviewed in section 3. During ablation procedures, different lesion patterns are created. Circular lesions are used to isolate areas of potential triggers for AF. Additional linear lesions are supposed to prevent reentry paths. Substrate based, point like ablation lesions are supposed to destroy substrates maintaining AF, such as zones of slow conduction [2].

3

Radiofrequency Ablation of Cardiac Tissue

3.1 Basics of Radio Frequency Ablation

The general principle of Radiofrequency Ablation (RFA) is a local tissue heating due to high frequency currents. The frequency range is 300-1000 kHz. One prerequisite is a frequency well above 100 kHz to avoid any stimulation of cardiac or other excitable cells [4]. Commonly RF-currents are applied in a unipolar configuration. Current flows from the catheter tip electrode to a larger reference electrode (dispersive electrode) at the body surface. Thereby, a high current density at the tip of the catheter is created which leads to resistive heating. Current density diminishes within a distance of several millimeters. Therefore, convective heating is the predominant mechanism in deeper tissue layers [35]. Maximum lesion depth is reached after 45-60 s of continuous RF current flow [36]. Permanent tissue damage is reached at temperatures between 50 °C and 60 °C [36], [37]. Lesion size is influenced by a number of constraints, which are listed in Tab. 3.1.

3.2 Lesion Morphology and Size

The core of an RFA lesion is described to consist of necrotic tissue. Moreover, an optically invisible border zone of up to 6 mm from the lesion edge can still show micro-structural injury [40]. Ndrepa and Estner described two zones: A necrotic core and a hemorrhagic border zone [44].

Table 3.1. Influencing parameters for RFA lesion size and depth

Electrode Radius	For larger electrodes, a linear increase of lesion width with increasing radius was found [4, 38] .
Electrode length	Electrode lengths between 4 mm and 10 mm produced a linear increase of lesion size and width [39].
Peak temperature	Lesion depth increased linearly with increasing maximum temperature measured at the catheter tip [38, 40, 41].
Maximum current	A linear relationship of maximum applied current and lesion depth was measured [38].
Maximum power	Lesion depth was found to depend on the maximum applied power in a linear way [38, 41, 42].
Application time	Lesion depth increased logarithmically over time, reaching a steady state after 45-60 seconds [40, 42].
Contact force	Greater contact forces led to a linear increase of lesion width [43]

Lesion size is influenced by the heat capacity of the tissue surrounding the lesion. Circulating blood in the chamber cools the endocardial surface, which prevents a large extent of hyperthermia at the endocardial surface. In the midmyocardial region this effect is not present, which leads to a greater diameter of the lesion. In epicardial direction, the lesion narrows to form an ellipsoidal shape. Volume loss in ablated tissue may lead to an indented endocardial scar surface [44]. However, the opposite effect may occur by inflammatory swelling. If temperatures exceed 100 °C, tissue at the surface can be vaporized.

Assuming an ellipsoidal shape of the lesion, it can be described by its width and depth. These parameters have been studied in various experiments (Tab. 3.2).

Table 3.2. Overview of lesion dimensions found in literature; lesion depth and width in mm

Study description	lesion depth	lesion width	ratio
Bovine liver in-vitro, different superfustion flow rates [45]	5.0	8.3	1.7
Bovine liver in-vitro, different maximum temperatures [46]	6.1	10.9	1.8
Sheep atrium in-vivo [47]	2.2	5.3	2.4
Sheep ventricle, in-vivo [47]	4.2	5.9	1.4
Rabbit, ventricle, in-vitro [47]	2.8	5.4	1.9
Pig atrium, in-vivo, orthogonal catheter orientation [6]	3.2	4	1.3
Pig, atrium, in-vivo, parallel catheter orientation [6]	3.2	5.6	1.6

3.3 Electrophysiological Changes Caused by RF Currents

Temperature alterations and current flow during RFA are assumed to cause various changes of cellular electrophysiology. Several studies dealt with the influence of hyperthermia on cardiac tissue. Data supports the assumption, that this is the predominant reason for myocardial injury [48]. Heated solution was used to produce states of elevated temperature in cardiac tissue [36, 49, 50]. This experiment enabled continuous electrical recordings and ensured controlled, homogenous tissue heating. A study by Simmers et al. aimed at translating these results to radiofrequency based measurements [51]. Few studies presented micro electrode recordings directly after RFA [52, 53]. However, RF-currents prevent electrical measurements during ablation. A study by Wu et al. overcame this limitation by using voltage sensitive optical mapping and a fiberoptic probe to measure the action potential duration during RFA [54].

3.3.1 Cellular Effects Related to Hyperthermia Due to Solution Based Tissue Heating

Nath et al. investigated electrophysiological parameters of guinea pig papillary muscles by exposing them to solutions at temperatures in the range of 38 - 56 °C. Changes in transmembrane voltage (ΔV_m), maximum action potential (AP) slope steepness ($\Delta dV/dt_{max}$), AP amplitude (Δ AP amplitude) and AP duration (Δ APD) were recorded. As a conclusion of these results, three temperature states were distinguished:

1. 38 - 45 °C: Normal function, higher dV/dt_{max}, AP shortening, smaller AP amplitude
2. 45 - 50 °C: Progressive depolarization of the sarcomere, distinct AP shortening, amplitude decrease
3. above 50 °C: Partial or completely irreversible contracture, permanent depolarization, cells are unexcitable, conduction block

The described temperature ranges for conduction block were partially confirmed by Simmers et al. [50]. They recorded CV, $\Delta dV/dt_{max}$, and loss of conduction on an isolated canine ventricular preparation. They found the temperature for reversible conduction block to be 50.3 $\pm$ 1.1 °C and for irreversible block 53.6 $\pm$ 0.6 °C. However their findings regarding ($\Delta dV/dt_{max}$) contradict the study of Nath el al. as they found a decrease above 45.5 °C. Detailed measurements of CV were performed in these experiments. A baseline CV of 0.35 $\pm$ 0.13 m/s was reported. CV was above baseline level for temperatures 38.5 - 45.4 °C with a maximum at 41.5 - 42.5 °C with an increase up to 114 %. Above 45.5 °C CV fell below baseline CV and further decreased, till conduction block occurred.

3.3.2 Analyses of Electrophysiological Parameters Based on Electrical Measurements After RFA

The emerging question from the previous section is if these results can be directly related to temperature rises caused by RF-currents. To evaluate this, Simmers et al. used the same dog model as in [50], but included an ablation electrode and a thermocouple for temperature measurement within the tissue [51]. The thermo couple was positioned at 200 μm underneath the ablation electrode. Using this setup they found reversible conduction block to occur at 50.7±3.0 °C and irreversible block at 58.0±3.4 °C, which is significantly higher than the temperature found using heated superfusate. However, they stated, that this difference results from varying temperature rise times and the position of the thermocouple, leading to an overestimation of the real temperature necessary for conduction block.

Two studies conducted micro electrode recordings before and after RFA [52], [53]. Ge et al. acquired APs in an area around 1-12 mm from the lesion edge in a time interval from 1-30 min post ablation. Superfused canine ventricular strips (2-4 x 2-3 x 0.2-0.3 cm) served as a tissue model.
The major findings of these studies for the border zone of RFA-lesions are:

Table 3.3. Summary of AP-changes as documented by Wood and Fuller (WF) and Ge et al. (GE); table shows maximum changes (%)/ spread of changes from lesion edge (mm)

	GE	**WF**
APD	-57 % / 8 mm	-19 % / 2.5 mm
Amplitude	-61 % / 6 mm	-16 % / 0.25 mm
Vm	-39 % / 4 mm	not investigated
dV/dt	-78 % / 6 mm	unchanged
CV	not investigated	23 %

1. APD is shortened
2. AP amplitude is reduced
3. Resting membrane voltage is higher (depolarization)
4. Conduction velocity is altered

However, the extent of the border zone differs largely. Ge et al. found myocytes to be affected up to 8 mm from the lesion edge, whereas the effects Wood and Fuller reported, are limited to a range of up to 2.5 mm. The used animal model may serve as a plausible explanation for these differing results. Wood and Fuller used a perfused whole heart, whereas Ge et al. ablated ventricular strips kept in a superfusion bath. Therefore cooling effects and a bigger thermal capacity in the whole heart may limit the RFA related effects and lead to a more realistic model regarding their spatial extent. Additional information on tissue related changes is given in a study of perfusion conditions in acute RFA lesion areas [55]. It was found, that these areas are less perfused after RF application.

3.3.3 Effects Occurring During Ablation

During RFA electrical measurements are disturbed by ablation currents. Therefore, APD measurements under these conditions can only be performed using optical mapping and voltage sensitive dyes. Wu et al. [54] analyzed APD_{80} with a fiberoptic fluorymeter at several points adjacent to the lesion edge (distance 0 - 6 mm) in rabbit Langendorff-perfused hearts. Due to bleaching effects AP amplitude cannot be measured by optical means. They observed a rapid decrease of APD_{80} after the onset of RF-current, which recovered partially within 15 min after the RF pulse. The results of this study suggest, that there are fast mechanisms, which affect APD but cannot be related to temperature rise. APD shortening is reported to start at 1 s after RF onset. Whereas temperature takes 15 s of ablation to reach $47°C$ [51], which would lead to the reported changes. Experiments of this study were performed at room temperature. This significantly affects baseline APD (compare $APD_{90}=130$ ms, Wood & Fuller and $APD_{80}=350$ ms, Wu et al., both in rabbit). No follow up studies have been conducted to evaluate the results of Wu et al. therefore, until today no explanation exists for the fast changes at the onset of RFA.

3.3.4 Long Term Development (1 Day to Weeks Post Ablation)

Wood and Fuller [52] performed open chest epicardial ablations in a rabbit model. Hearts were extracted 1 week and 1 month post ablation and micro electrode measurements were performed close to lesions and in vital tissue. In this study no significant changes of APs in the border zone could be found in comparison to reference measurements.

3.4 Intracellular Ion Concentrations During Hyperthermia

Everet et al. [56] studied calcium release during hyperthermia in guinea pig papillary muscles. They reported a significant increase of intracellular calcium after a 60 s exposure to 48 °C solution compared to 42 °C. Using specific blockers they were able to rule out temperature influences on calcium channels. Therefore, they concluded, that calcium rise occurs due to a channel independent increase in calcium permeability of the membrane [56]. As measurements were performed by fluorescence optical mapping, no quantification of intracellular calcium level was possible.

A second study investigated intracellular potassium in liver tumor cells [57]. They reported a slight increase after 20 min of heat treatment followed by a more pronounced decrease up to 120 min. However, time development of this process is too slow to come into play during RFA.

4

Intracardiac Electrogram Acquisition and Analysis

Intracardiac Electrograms (IEGMs) are clinically measured inside the heart using measurement electrodes attached to catheters. This minimally invasive technique is important for the diagnosis and treatment of cardiac arrhythmias. During an electrophysiological procedure, the physician inserts a catheter via a small cut into the femoral vein at the left leg into the venous system. Following the veins the catheter is forwarded into the right atrium under fluoroscopic guidance [4]. For access to the left atrium, the septum is punctured. Catheter electrodes transform ionic current flow into an electronic current, which can be measured by a signal acquisition system. Alterations of the extracellular potential due to the measurement system are minimized by high input impedance, which prevents current flow. Clinical electrodes are usually made from stainless steel with platinum coating [58].

4.1 Underlying Mechanisms of Extracellular Potentials

When a wave front propagates over a piece of myocardium, sodium influx into the cells causes an intracellular current in the direction of propagation. Extracellularly, a compensating current flow in the opposite direction is established. Thus, an extracellular potential distribution is generated [59]. These phenomena are visualized by the simulation data shown in Fig. 4.1. In the intracellular domain (Fig. 4.1(a)) the wavefront, which travels from left to right, leads to a depolarization of the myocardium in the back of the wavefront. From this positively charged part, current flows to the area on the

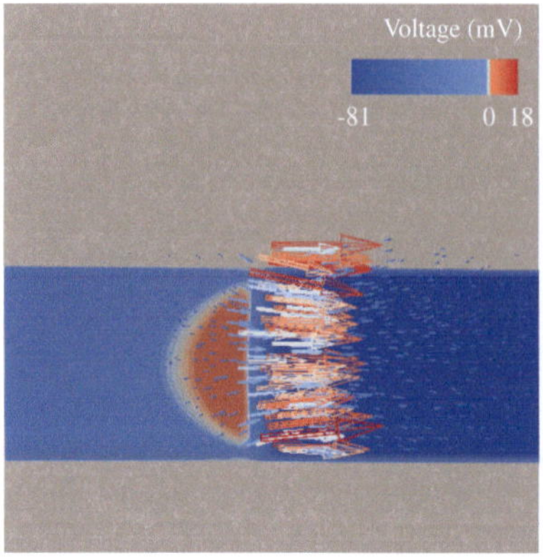

(a) Transmembrane voltage and intracellular currents

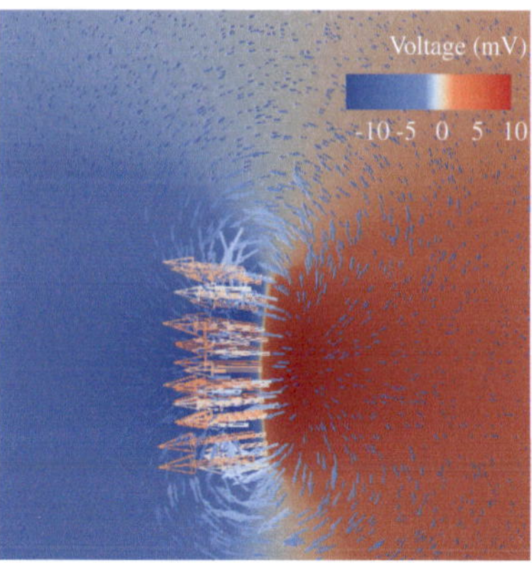

(b) Extracellular potentials and extracellular currents

Figure 4.1. Current and voltage distributions around a cardiac depolarization front. Currents are displayed by colored arrows pointing in the direction of current flow. Arrow colors and length represent the magnitude of the current density. The starting point of an arrow coincides with the point, where the current is measured. (a) Intracellular currents and the corresponding transmembrane voltages. Although currents were calculated based on intracellular potentials, transmembrane voltages are shown, as they are more meaningful. (b) Extracellular potentials and extracellular current flow.

side, which is still in rest. The highest current density is present at the excitation front. Currents from and to the surrounding tissue are small. Outside the cells (Fig. 4.1(b)), in the extracellular space, the current loop is closed by a backwards directed current. Due to this current dipole, an extracellular wavefront is created, which provides a positive electric field component at the front and a negative in the back. Using electrodes these extracellular potentials can be measured as voltage between two electrodes.

4.1.1 Definition of Unipolar and Bipolar Electrograms

Electrical voltages are measured as difference of two local potentials. For the measurement of extracellular potentials, two configurations are defined. Unipolar EGMs (UEGM) are measured between one point at the cardiac surface and a reference electrode, which is ideally placed at infinity. In clinical practice this electrode is usually a patch electrode at the body surface or the so called Wilson's central terminal (WCT). The WCT is defined as the mean of the two arm and the left leg potentials (Fig. 4.2) [59, 60]. Two exemplary simulated UEGMs are shown in figure 4.3(a). An approaching wavefront creates a positive potential on the measuring electrode, which is the so called R-wave of the unipolar EGM. When the wavefront is directly underneath the electrode, the negative deflection is formed, which is associated with the local activation time. The following negative S-wave results from the negative back of the wavefront [58].

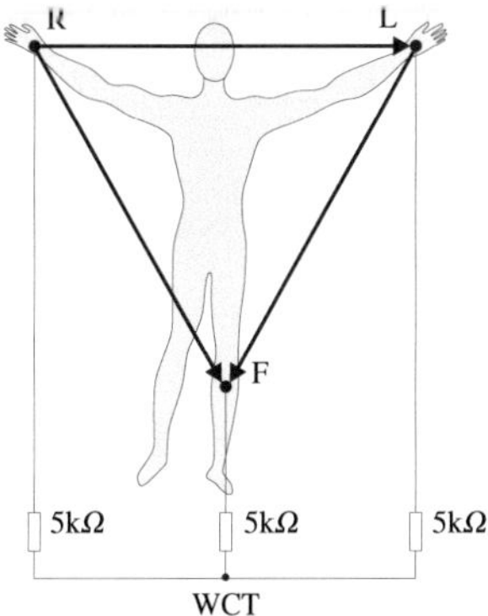

Figure 4.2. Wilson's central terminal lead is measured between the two arms and the left leg. All leads are connected over 5 kΩ resistors, adapted from [12].

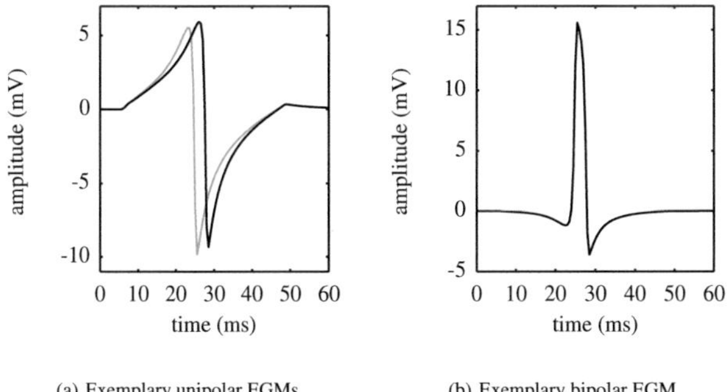

(a) Exemplary unipolar EGMs (b) Exemplary bipolar EGM

Figure 4.3. Exemplary intracardiac EGMs; (a) Two unipolar electrograms from locations 3 mm distant from each other, both touching the endocardium. (b) Resulting bipolar EGM representing the difference of the signals in (a)

Bipolar EGMs (Fig. 4.3(b)) are measured by two electrodes near the myocardium, which usually have a spacing of a few millimeters. This configuration is advantageous due to the cancellation of far fields, like the electrical mains or ventricular signals during measurements in the atria. Also, the field of view is limited to the local electrical activity in the vicinity of the electrodes. As bipolar EGMs are dependent on the electrode orientation, caution is advised for the interpretation of bipolar EGM morphology and amplitude [58]. In today's clinical practice, bipolar recordings are the gold standard in atrial mapping procedures. However, recent advances have shown advantages of using unipolar EGMs e.g. in the annotation of local activation times [61].

4.2 Catheters in Clinical Use

Various catheters are used in clinical practice to measure extracellular potentials (Fig. 4.4). Depending on the intended application, electrode configuration, number and size is varied. Electrodes are usually made of stainless steel or platinum [4, 58]. For sequential mapping of the endocardial surface, catheters with about 20 ring electrodes are used. Geometries vary between circular and finger like structures. A catheter, which can cover large portions of the atrial wall is the so called basket catheter. It provides 128 mapping electrodes. For ablation, catheters with larger electrodes are used in order to receive a larger contact area with the endocardial tissue for RF current delivery. Some ablation catheters have a perforated tip electrode, which can be perfused by saline to cool the surface area during ablation. Signals from catheter electrodes are acquired and digitized by clinical recording systems offered by several companies. Some systems also offer a 3D-localization of the electrodes inside the heart. This is either done by magnetic field sensors included in the catheter tip or by electric impedance measurements between the catheter electrodes and three reference electrodes at the body surface of the patient [4].

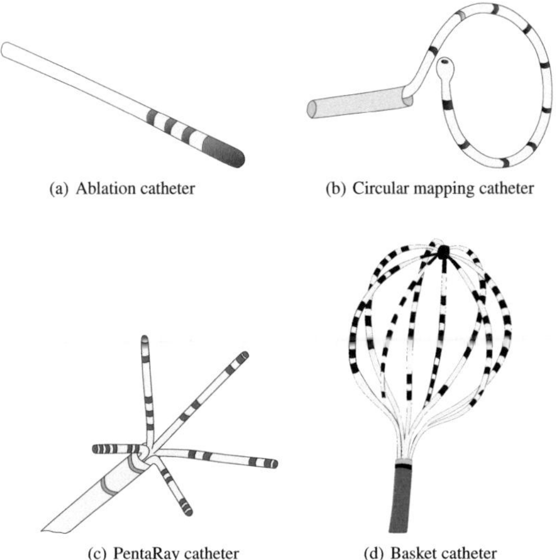

Figure 4.4. Different catheter types used in today's clinical procedures

4.3 Segmentation of Intracardiac Electrograms

In this work an algorithm for automatic electrogram segmentation was used, which was presented in [62] and [63]. In order to find electrograms and electrically active segments in a measured signal, several processing steps are performed.

Measured signals are filtered to remove baseline wander and high frequency noise. Using the non linear energy operator (NLEO) [64], the temporal energy of the signal is calculated. The NLEO of a discrete signal x_n is defined as:

$$NLEO_n = x_n^2 - x_{n+1} \cdot x_{n-1} \qquad (4.1)$$

The resulting signal has an amplitude, which is proportional to the square of the amplitude of x_n and to the square of the frequency of x_n. In order to receive a hull curve of this signal, a Gaussian low pass with a cut of frequency at 24 Hz is applied. Subsequently active segments are identified by adaptive thresholding, which accounts for temporally variable amplitude levels. In a post processing step, very short active and inactive segments are removed.

In [62] and [65] an additional method was proposed to identify those active segments, which fit to a detected dominant frequency in the signal. In a first step, all characteristic frequencies within a range of 4-10 Hz in the power density spectrum are identified. For each detected characteristic frequency, active segments are extracted, which fit the corresponding cycle length. The result is an assignment of each detected EGM to one or more characteristic frequencies. Assuming that one trigger source produces activations with a constant delay, all activations, which belong to this trigger, can be found.

4.4 Complex Fractionated Atrial Electrograms (CFAE)

In diseased hearts, intracardiac electrogram shape often looses its physiological shape and displays an increased number of deflections per activation. Areas, where these complex fractionated atrial electrograms (CFAE) are measured turned out to be a suitable target for RF-ablation [5]. CFAEs were defined as bipolar EGMs which display more than one deflection, a permanently disturbed baseline or a basic cycle length between two activations $< 120\,\text{ms}$ [5]. It was found, that the number of CFAE sites in hearts with atrial fibrillation is increased, as well as the volume fraction of fibrosis [23]. This suggested the hypothesis, that CFAEs originate from fibrotic areas. The labyrinth-like propagation patterns in these areas are assumed to cause electrogram fractionation. However, there are more possible reasons for CFAEs, which also include remote activation from adjacent structures, fibrillatory excitation patterns, or filter artefacts [59]. Therefore, the underlying mechanisms for CFAEs are a topic for ongoing scientific discussion [66].

4.4.1 Database of Complex Fractionated Electrograms

In section 13.5 simulated electrograms are compared to a database of clinical CFAE signals. This database was presented in [62]. Signal pieces with

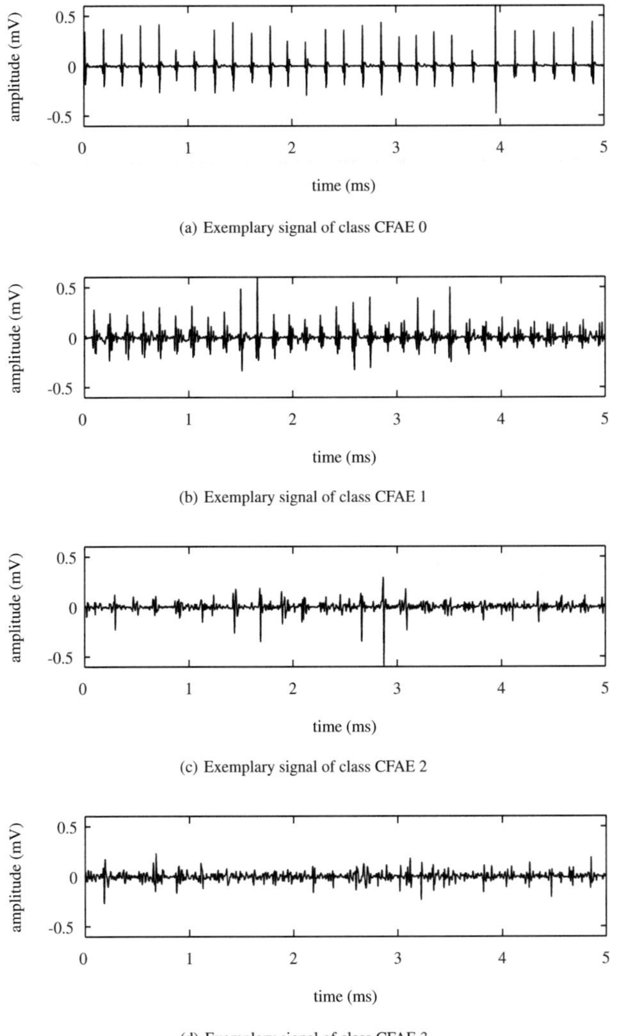

(a) Exemplary signal of class CFAE 0

(b) Exemplary signal of class CFAE 1

(c) Exemplary signal of class CFAE 2

(d) Exemplary signal of class CFAE 3

Figure 4.5. Examples for the four signal classes in the CFAE database presented by [62].

a length of 5 s, recorded with circular mapping catheters, were annotated by two electrophysiologists from independent centers based on a defined nomenclature. The annotation reaches from class CFAE 0 for EGMs without any fractionation up to class CFAE 3 with continuous activation. Exemplary signals for each class are shown in Fig. 4.5. CFAE nomenclature was defined based on clinical observations and the established pattern used by electrophysiologists in their search for ablation targets. Class CFAE 0 includes all types of non-fractionated signals independent of the basic cycle length. Clearly distinct EGMs with a similar pattern of fractionation belong to class CFAE 1. If the measured pattern of fractionation is variable over the 5 s recordings, signals are classified as CFAE 2. And finally, continuous activity is described by class CFAE 3.

Experimental Studies of Cardiac Electrophysiology

The heart of mammals is a complex organ which is studied to gain new insights into cardiac pathologies. Experimental studies of cardiac pathologies reach from isolated single cells [67], cultured cells [68] up to whole heart preparations of large animals [69] and rejected human donor hearts [70]. Measuring modalities of cardiac electrophysiology include microelectrode measurements of transmembrane voltages, measurements of extracellular potentials and fluorescence optical measurements of transmembrane voltage or cytosolic calcium. In this chapter a short introduction is given into the optical measurement of transmembrane voltages as well as the acquisition of extracellular signals from the surface of tissue preparations.

5.1 Optical Measurement of Transmembrane Voltages

The measurement of transmembrane voltages using fluorescent dyes is commonly called "optical mapping" [71]. This technique allows to measure the electric activity of myocardial tissue in a contactless way with high spatial and temporal resolution. Depending on optical magnification the spectrum reaches from the sub-cellular to the whole organ level [72–76].

5.1.1 Fluorescent Dyes

A variety of fluorescent molecule classes offer the ability to indicate the transmembrane voltage. Underlying mechanisms include changes in molecule orientation, distribution and activity. However, time constants in these processes are too long to serve as marker for the fast changing transmembrane voltage of cardiac cells [76]. The only mechanism capable of following the fast upstroke of an AP is a charge shift within the molecule under the influence of the changing transmembrane voltage. This mechanism is called a molecular Stark effect [77] or electrochromism [78]. The response time of these fluorescent dyes lies in the nanosecond range [79]. The most frequently used voltage sensitive dye (VSD) is called di-4-ANEPPS. This VSD is excited by light in the green range of the visible spectrum. It reaches a higher energetic level by a conformation change within the molecule, which is mainly expressed by a charge shift along the long axis of the molecule [79] (Fig 5.1). The ability to indicate the transmembrane voltage depends on the

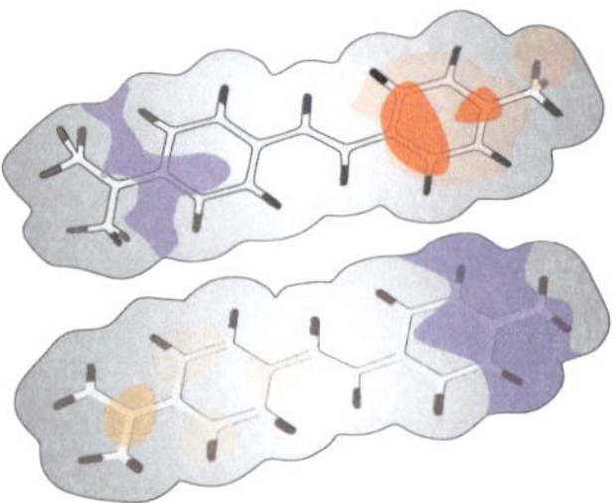

Figure 5.1. Charge distribution along the axis of the di-4-ANEPPS molecule. The molecule at the top shows the excited state of the molecule. The molecule at the bottom illustrates the ground state. Red areas show electron donors, blue areas show electron acceptors. During the transition from the ground state to the excited state, electrons are shifted along the molecule axis. Image modified from [79].

orientation of the dye molecule relative to the cell membrane. The dye has one lipophilic and one hydrophilic end. Therefore, it always attaches to the outside of the cell membrane in a defined orientation. If the electric field across the membrane changes, the charge shift is supported by the electric field, which leads to a blue shift to higher energy levels in the excitation and emission spectrum [77] (Fig. 5.2). Due to the fact, that the shift of the gaussian shaped spectra is in the range of 2 nm it can hardly be detected by a color change. Therefore, an intensity based approach is used to measure the course of the transmembrane voltage. In order to achieve a high sensitivity, the dye is excited by a narrow bandwidth light source at a steep falling part of the excitation spectrum (e.g. 520-540 nm). From the emission spectrum also the long pass part (e.g. >610 nm) is extracted. If the spectrum is shifted towards blue wavelengths, the excitation as well as the emission intensity decrease in the regarded parts of the spectrum. The measured intensity has a linear relationship to the transmembrane voltage [71] (Fig. 5.2). In recent years, derivatives of di-4-ANEPPS have been presented. Di-8-ANEPPS is optimized for single cell recordings, offering longer recording times due to slower wash out and internalization into the cell. The drawback of di-8-ANEPPS is its weak solubility in aqueous solutions [76]. Furthermore, dyes which emit in the infrared spectrum were used to perform transmural

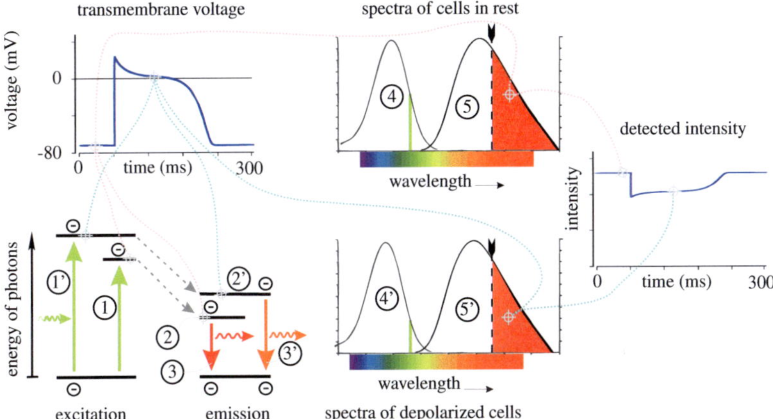

Figure 5.2. Simplified schematic of the transmembrane voltage response of the dye di-4-ANEPPS. The energy levels of the dye molecule, which is attached to the cell membrane are influenced by the transmembrane voltage. Under resting membrane conditions, the dye absorbs a certain percentage of the excitation light (1). This transfers the dye molecule into the excited state (2) from which it emits light of a certain wavelength (3). The quantity of absorbed and emitted light is governed by the excitation and emission spectra (4), (5). A common setting is, to excite the dye in the falling part of the excitation spectrum (4) and at the same time measure the intensity of the long pass part of the emission spectrum (5). The integral of the regarded longpass part of the spectrum leads to a detected intensity in the recording device. If the cell depolarizes, the changed transmembrane voltage influences the energy levels of the dye molecule (1'), (2'). This causes a blue shift of excitation and emission spectra (4'), (5'). Due to the shift, a reduced amount of the excitation light is absorbed (4'). Additionally, a smaller fraction of the emission spectrum contributes to the recorded longpass filtered wavelength range (5'). These effects lead to a drop in the intensity, which is acquired by the recording system.

recordings [80]. The dye ANNINE-6-plus is supposed to be a better version of di-4-ANEPPS with similar solubility, but a more pronounced change in fluorescence intensity due to membrane voltage changes [81].

5.1.2 Optical Mapping Systems

The basic optical components of an experimental system for optical mapping include an excitation light source, optical filters, lenses and a highly sensitive recording device (e.g. a quantum efficiency >70%) [71, 82]. Possible systems for optical data acquisition are photodiode arrays, laser scanners, CCD cameras or CMOS cameras [76]. The tradeoff between the different devices is the spatial and temporal resolution. Photodiode arrays offer a high temporal resolution, which only depends on the AD-conversion frequency of the analog diode signal. However, these devices are often limited to a 16×16 array. Highly sensitive CCD cameras offer a higher spatial resolution, but their temporal resolution is limited to about 1 kHz. In the past years CMOS cameras have reached the necessary sensitivity [83] and will probably be the best choice for future recording systems.

Several light sources can be used for excitation of the fluorescent dye. The first used sources were tungsten or halogen lamps together with an optical filter to choose the specific excitation wavelength. Today lasers and light emitting diodes are used [71]. In order to extract the relevant light spectrum for excitation and emission a suitable setting of optical filters has to be used. High sensitivity for di-4-ANEPPS can be achieved by an excitation between 500 and 540 nm and recording of wavelengths above 600 nm [77]. In the case of epi-illumination an additional dichroic mirror is necessary to couple the light into the light path of the recording system. In this approach, the excitation light reaches the preparation in the region of interest at the same angle at which the recording takes place. This illumination technique was shown to produce a better signal to noise ratio in recorded optical signals [76].

Emitted light needs to be collected and focused. For this purpose either single lens, double lens systems or for high magnification microscopes can be used. It has been shown, that regarding the acquired light intensity double lens systems are more efficient [76]. Moreover for low magnification ($<5\times$) the so called "tandem lens macroscopes" have a better light collecting ability than conventional microscopes [84]. It was shown, that achieved image brightness of these low magnification systems can reach the 100 fold value of a microscope [85]. A version of a tandem lens macroscope was used in the experimental setup presented in this work and is shown in Fig. 7.6.

5.1.3 Image Acquisition Approaches for Frame Rate Interpolation

Frame rate for the acquisition of optical action potentials and excitation propagation needs to be well above 700 Hz and ideally in the range of 1 kHz [86]. If the used camera cannot deliver this high frame rate a method for frame rate interpolation can be used. The idea behind this interpolation is, that a stimulated preparation repetetively shows the same excitation pattern relative to each stimulus. By recording several image sequences, which are shifted relative to the stimulus, an interpolated frame rate can be achieved (boxcar averaging in waveform recovery mode [87]). Based on the algorithms presented in [88] and [89], a recording scheme for frame rate interpolation was implemented in the experimental setup presented in chapter 7.

5.1.4 Analysis of Optical Mapping Data

Due to the small fractional change in intensity, optical mapping data is always subject to low signal to noise levels. Therefore, several steps of preprocessing have been established. Most analysis algorithms include the following steps [76, 83]:

- Choice of a region of interest by manual or automatic segmentation
- Spatial filtering to remove noise within one time frame
- Temporal filtering to remove noise in the time course of intensity for each pixel

- Drift removal caused by optical bleaching or motion
- Normalization of the intensity per pixel in order to account for inhomogenous dye distribution

Each of these steps has to be adjusted to fit the specific needs of the study. Therefore, a variety of workflows can be found in literature.

5.2 Measurement of Extracellular Potentials

Electrical processes in a biological environment are mostly based on ionic conduction. Therefore, the first task in their measurement is a translation of ionic currents into electronic currents, which can be quantified by conventional electronic circuits. The interface in between these domains is formed by a measurement electrode. Subsequently, measured voltages need to be amplified and converted to digital values [58, 90, 91].

5.2.1 Electrode Materials

At the electrolyte-electrode interface, charge transfer can be based on different mechanisms. If no chemical reactions take place, the charge transfer is purely capacitive. Such an electrode would be called "polarized electrode". The other major effect are redox reactions between ions and the metal. These reactions produce a so called faradaic current over the interface [91]. Further charge transfer is possible due to adsorption of ions or other side effects due to chemical reactions [90].

Materials used for electrophysiological measurements are usually made of so called noble metals such as silver, gold or platinum [90]. In some cases also stainless steal is used. Their common property is, that they are not easily oxidized in salt based solutions. These electrodes belong to the category of polarized electrodes, which have apart from a very small leakage current a capacitive characteristic. A simple equivalent circuit of a polarized electrode is a capacitance with a high resistor in parallel (Fig. 5.3).

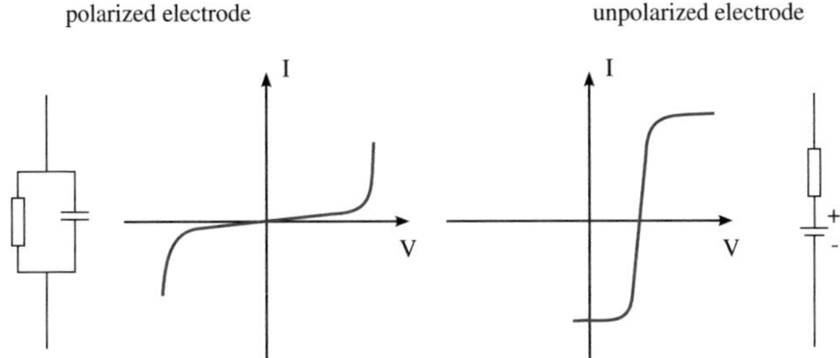

Figure 5.3. Idealized electrode types modified from [90]

More complex equivalent circuits for the case of platinum electrodes are described in [92]. Another frequently used electrode is the silver/silverchloride electrode (Ag/AgCl). This electrode is a so called unpolarized electrode (Fig. 5.3). If an electric field across the electrode/electrolyte interface is present, this immediately leads to a faradaic current due to the low charge transfer resistance [93]. This effect can be explained by the AgCl layer at the surface of the Ag electrode [91]. It facilitates the redox reaction between the chloride containing electrolyte and the electrode material [90]. Amplifiers with high input impedance reduce the current nearly to zero.

5.2.2 Amplification and A/D-Conversion

Extracellular signals recorded from cardiac preparations are in the millivolt range. It has to be ensured, that the cells, which serve as sources are not exposed to an electric load. Due to the impedance of the source the measured voltage would drop significantly. Therefore, the measurement system needs to have a high input impedance. Measured signals need to be amplified to fit the measurement ranges of suitable A/D-converters. For differential measurements, usually the first stage is formed by an instrumentation amplifier. Besides a first amplification of the differential signal, this circuit provides an input impedance above $10^8 \Omega$. According to the constraints of the following A/D-conversion system, additional amplifiers are put in series to the input stage [90]. Sampling frequency of A/D-conversion in today's clinical systems ranges around 1 kHz. In experimental systems, in particular in the case of small spatial distances between measurement electrodes, sampling frequencies of 50-100 kHz are applied [94, 95]. Due to high noise levels, high amplitude resolution is of minor concern in the case of extracellular recordings. Therefore, usually precisions of 12-14 bit are sufficient.

5.3 In-Vitro Experiments on Vital Cardiac Tissue

Isolated preparations of cardiac tissue form the basis for in vitro-experiments on cardiac electrophysiology. Animals are anesthetized with an overdose of narcotics. Then, the heart is rapidly excised and placed into nutrient solution, keeping the tissue vital. Different approaches exist to keep the tissue alive throughout the experiment, which can last for several hours.

Whole heart preparations can be supplied by perfusion. This implies the use of coronary vessels, which originate at the aortic side of the aortic valve. Most commonly for this purpose a so called Langendorff perfusion is introduced. The aorta is cannulated and nutrient solution is pushed through the cannula in a constant pressure or constant flow mode. The flaps of the aortic valve close and the perfusate is pushed through the coronary arteries. After the passage of the coronary system, the fluid is collected in the coronary sinus, which drains into the right atrium [84].

It is also possible to dissect parts of the heart. These preparations can also be supplied by perfusion, if coronary vessels exist, which are wide enough to be cannulated. In particular for small animal hearts this is not possible. If the preparation is thin enough (<1 mm), it can be supplied by superfusion, whereas deeper tissue layers are reached by diffusion [84]. An autorhythmic superfused preparation of the right atrium is presented in [96, 97]. Also the papillary muscles are a frequently studied preparation [49, 84, 98].

5.3.1 Nutrient Solutions

Vital cardiac tissue preparations need to be kept under constant physiological conditions. This includes physiological ion concentrations of the relevant ions (Na^{2+}, K^+, Ca^{2+}, Cl^-, Mg^{2+}), glucose as a source of energy, as well as oxygen. Furthermore, a pH-value of 7.4 needs to be reached. These constraints are fulfilled by solutions as Krebs Henseleit solution (KHS) or Tyrode solution. The solution is gassed with carbogen gas (95 % O_2, 5 % CO_2), which delivers a high partial oxygen pressure. The carbon dioxide is necessary for pH adjustment [84].

6

Simulations of Cardiac Electrophysiology

Myocardial tissue is formed by cells with average dimensions of $100\,\mu$m in length and $10\,\mu$m in width [12]. These cells are aligned along a dominating fiber direction. Zooming in to the cellular level, each cell has several thousands of ionic channels, which form the basis for cardiac electrophysiology. One ionic channel is formed by several protein subunits, which interact with each other and the adjacent environment to perform their specific tasks [99]. Each of these levels can be described in detail by computational models. However, the computational cost for a whole heart simulation accounting for all these features would need an enormous number of nodes in a super-computing cluster or would take years to compute on a high performance computer with several dozens of nodes. Fortunately, for each specific aim of a simulation study already models exist, which are capable to represent the parameters of interest at an affordable computational cost. This is possible due to carefully chosen simplifications. In this work, extracellular potentials and currents on the tissue level are studied. A model of high accuracy on this scale is the so called bidomain model [100–102], which is described in section 6.2.1. This model uses a diffusion based description of the ionic currents. The cellular activity is described by a cellular model for each computational element. A cellular model is based on differential equations describing the ionic current flow for each type of ionic channels. For the simulations presented in this work the model according to Courtemanche et al. for human atrial cells [103] was used. The behavior of ventricular myocytes was represented by the model according to ten Tusscher et al. [104] and a version of this model accounting for ischemia related effects by Wilhelms et al. [105].

6.1 Cellmodels

In today's models for cellular electrophysiology, ion currents through the cell membrane are approximated by a set of linear differential equations. This idea was first introduced by Hodgkin and Huxley [106]. In this concept, the current contribution of each active cell membrane protein type x is reflected by an equation of the type:

$$I_x = g_x(V_m - E_x) \tag{6.1}$$

In this equation I_x is the specific ionic current, g_x is the conductivity of the membrane protein and E_x is the Nernst potential for the relevant type of ions transported by x. The dependency of the membrane protein conductivity on time and V_m is reflected in the formulations of g_x, which includes the maximum conductivity of the proteins $g_{x,max}$ and the product of several gating variables γ_i

$$g_x = g_{x,max} \cdot \prod_i \gamma_i. \tag{6.2}$$

In case of the fast sodium channel in the Hodgkin Huxley model [106], g_{Na} depends on two gating variables. The activation is represented by the gating variable m and the inactivation by the gating variable h. Sodium conductivity can be written as:

$$g_{Na} = m^3 h \cdot g_{Na,max} \tag{6.3}$$

The dependencies of the gating variables m and h on transmembrane voltage V_m are reproduced by linear differential equations:

$$\frac{dm}{dt} = \alpha_m(1 - m) - \beta_m \tag{6.4}$$

$$\frac{dh}{dt} = \alpha_h(1-h) - \beta_h \tag{6.5}$$

Where α_x and β_x are the functions of V_m. Fulfilling the boundary conditions $m = m_0$ and $h = h_0$ for $t = 0$, equations 6.4 and 6.5 are solved by:

$$m = m_\infty - (m_\infty - m_0)e^{-t/\tau_m} \tag{6.6}$$

$$h = h_\infty - (h_\infty - h_0)e^{-t/\tau_h} \tag{6.7}$$

Where $m_\infty = \alpha_m/(\alpha_m + \beta_m)$, $\tau_m = 1/(\alpha_m + \beta_m)$ and $h_\infty = \alpha_h/(\alpha_h + \beta_h)$, $\tau_h = 1/(\alpha_h + \beta_h)$ [106]. Descriptions of other gating variables γ_i can be found in [12, 107].

The summation over all active membrane protein currents I_x provides the overall ionic current passing the membrane I_{ion}:

$$I_{ion} = \sum_x I_x \tag{6.8}$$

Transmembrane voltage V_m is affected by the transmembrane current (ionic currents I_{ion} and external stimulus current I_{stim}) and a capacitive component introduced by the membrane capacity C_m. Thereby, changes in transmembrane voltage can be described by the following differential equation:

$$\frac{dV_m}{dt} = -\frac{I_{ion} + I_{stim}}{C_m} \tag{6.9}$$

6.1.1 Human Atrial Cell Model According to Courtemanche et al.

Published in 1998, the atrial cell model by Courtemanche et al. [103] was the first cardiac atrial cell model adjusted to human data. In the following this model will be termed 'Courtemanche model'. It comprises 12 different types of ion channels and pumps. Specifically, these are one fast sodium channel, five potassium channels of different temporal characteristics during the AP and one calcium channel. Furthermore, it includes the sodium-potassium exchanger, a sodium-calcium exchanger as well as two background leakage currents (sodium, calcium). Additionally to the membrane kinetics, also a model for intracellular calcium handling was included. A detailed study of atrial cell models can be found in [108, 109].

6.1.2 Human Ventricular Cell Model by ten Tusscher et al. Accounting for Ischemia Related Effects

Action potential characteristics of human ventricular myocytes are described by the model of ten Tusscher et al. ('ten Tusscher model'). It was originally published in [110] and presented in a refined version in [104]. It accounts for the same membrane channel types as the Courtemanche model, however membrane kinetics were mainly adapted to data of human ventricular myocytes. The ten Tusscher model includes advanced calcium handling. Furthermore, it distinguishes between different types of ventricular myocytes according to their transmural location (endo-, mid-, or epicardial).

The ten Tusscher model was adapted to different phases of ischemia by Weiss et al. [111] and Wilhelms et al. [105]. The included effects are related to hyperkalemia, acidosis and hypoxia, which develop in these poorly perfused regions of the heart. The impact of each of the components is weighted by a so called zone factor, which allows to separate between border and central zones of the ischemic region. Due to these effects, AP characteristics change. APD shortens, AP amplitude is reduced, which goes along with a gradual depolarization of the cell membrane, and finally AP upstroke velocity is reduced.

6.2 Modeling Cardiac Electrophysiology on the Tissue Level

Cardiac excitation is passed on from cell to cell which, leads to a propagating wavefront over the heart. This behavior was modeled in microstructural models, which account for each cell as a separate compartment [112]. When regarding the macroscopic aspect of cardiac excitation, intra- and extracellular spaces can be considered as two continuous domains, which are separated by the cell membranes. Current flow through the cell membrane is enabled by membrane proteins (channels or pumps). For each domain a homogenous and anisotropic conductivity is defined. E.g. for the intracellular space this conductivity tensor accounts for the cytosolic conductivity of the cell compartment as well as gap junctional resistance. Ionic current flow can be described by two Poisson equations, which form the bidomain model of cardiac excitation [12, 100, 113]. In case of equal anisotropy ratios of extra- and intracellular conductivity, the two Poisson equations of the bidomain model can be translated to one single Poisson equation. This is referenced to as the monodomain equation. In this work, the calculation of cardiac excitation propagation was performed using the software package acCELLerate [114].

6.2.1 Bidomain Model

Ionic current flow in the bidomain model can be calculated based on the assumption of two continuous domains for intra- and extracellular space. For each domain a current flow is defined by a Poisson equation (6.10), (6.11) with the cell surface to volume ratio β, intra- and extracellular potentials Φ_i and Φ_e and the conductivity tensors σ_i and σ_e.

$$\nabla \cdot (\sigma_i \nabla \Phi_i) = \beta I_m - I_{si} \qquad (6.10)$$

$$\nabla \cdot (\sigma_e \nabla \Phi_e) = -\beta I_m - I_{se} \qquad (6.11)$$

Apart from external stimulus currents I_{si} and I_{se}, all currents which flow out of one domain enter the other. By adding up equations (6.10) and (6.11), the first part of the bidomain equations (6.12), can be derived using the definition of the transmembrane voltage $V_m = \Phi_i - \Phi_e$.

$$\nabla \cdot ((\sigma_i + \sigma_e)\nabla \Phi_e) = -\nabla \cdot (\sigma_i \nabla V_m) \qquad (6.12)$$

By adding the definition of the transmembrane current $I_m = C_m \frac{dV_m}{dt} + I_{ion}$ the second part of the bidomain equations (6.13) can be obtained from equation (6.10).

$$\nabla \cdot (\sigma_i \nabla V_m) + \nabla \cdot (\sigma_i \nabla \Phi_e) = \beta \left(C_m \frac{dV_m}{dt} + I_{ion} \right) - I_{si} \qquad (6.13)$$

For a detailed derivation of the bidomain equations see [12, 100, 101].

6.2.2 Monodomain Model and Forward Calculation of Extracellular Potentials

For many questions in simulation studies, the bidomain equations can be simplified to improve computational performance. Given the condition, that σ_i and σ_e can be expressed by $\sigma_i = \kappa \cdot \sigma_e$, where κ is a scalar value (equal anisotropy ratios), the bidomain equations can be simplified to the monodomain equation [115].

$$\nabla \cdot (\sigma_i \nabla V_m) = (\kappa + 1)\beta \left(C_m \frac{dV_m}{dt} + I_{ion} \right) = (\kappa + 1) \cdot I_m \qquad (6.14)$$

This approach makes sense, if complex or large geometries of myocardium need to be simulated. If the corresponding extracellular potentials are of interest, a forward calculation based on transmembrane source current density can be performed [116]. The received transmembrane current can be regarded as source currents for the extracellular space [100, 117], where ϕ_e can be calculated based on the Poisson equation (6.11).

Boundary Conditions

Boundary conditions have to be applied, in order to solve the described equations. The most important types of boundary conditions are Neumann and Dirichlet boundary conditions, which will be explained in the following. If both types of conditions are applied at one boundary, conditions are called 'mixed boundary conditions'. Regarding a differential equation $\Phi(\mathbf{r})$ and a boundary S Dirichlet boundary conditions are defined as:

$$\Phi(\mathbf{r}) = q(\mathbf{r}); \ \mathbf{r} \ on \ S \qquad (6.15)$$

The conditions postulate a certain value in the solution of the differential equation on the boundary. In this context, Neumann conditions are defined, with n being the normal vector of the boundary:

$$\frac{\partial \Phi(\mathbf{r})}{\partial n} = p(\mathbf{r}); \ \mathbf{r} \ on \ S \tag{6.16}$$

In the case of Neumann boundary conditions, the derivative of the function along the normal direction of the boundary is defined. If $q(\mathbf{r})$ or $p(\mathbf{r})$ equal zero, the boundary conditions are called homogenous compared to inhomogenous for other cases [118].

Part II

Methods

7

Development of an Experimental Setup for Simultaneous Optical and Electrical Measurement of Cardiac Electrophysiolgy

In this chapter an experimental setup for electrical and optical measurement of cardiac excitation is presented. The setup is capable to acquire optical measurements of transmembrane voltage by a voltage sensitive fluorescent dye (di-4-ANEPPS) and electrical measurements of extracellular potentials from the same preparation in a simultaneous fashion.

7.1 Experimental Setup Overview

The experimental setup can be split into four parts. Transmembrane voltage is recorded by the optical part, which was optimized for the voltage sensitive dye (di-4-ANEPPS). This part includes a laser as excitation light source, several lenses for beam forming and optical filters. Optical signals are recorded by a sensitive CCD camera.

For the acquisition of extracellular potentials a measurement sensor was developed, which can be positioned by a computer controlled micro manipulator. A first amplification of the electrical signals recorded by two platinum electrodes is performed close to the tip of the sensor. Data is acquired and digitized by a custom made microcontroller board.

Myocardial preparations of various geometries can be kept vital under stable temperature conditions by a custom made tissue bath. In this bath the preparation is being continuously superfused by temperature adjusted nutrient solution, which is driven by a peristaltic pump. Cardiac preparations are pinned on a silicone disk and stimulated by additionally positioned tungsten wires.

In order to achieve real time conditions, control of the measurement sequence is performed by a second micro controller, which solely controls the timing tasks. These tasks include the opening of the shutter, start and length of the stimulation pulse, triggering of the camera in a frame-by-frame mode as well as the start and duration of the electrical measurement.

7.1.1 Components of the Experimental Setup

The central part of the experimental setup is built inside a Faraday cage, which is formed by copper sheets attached to a wooden frame (Fig. 7.1(a)). The inside walls are covered by molleton, which is highly light absorbing (Fig. 7.1(b)). The center part of the experimental setup is built on a table with vibration isolating mounting feet. Except for the bottom plate of the Faraday cage this table is mechanically uncoupled from the Faraday cage. Along the four corners of the cage, steal bars are provided as a mounting base for components above the central part in a mechanically uncoupled way. (Fig. 7.1(c)). The central part of the setup (Fig. 7.2) is mounted on a $600\,\text{mm} \times 600\,\text{mm}$ breadboard (Thorlabs Inc.,Newton, NJ, USA). The majority of components belong to the optical part, which will be introduced in section 7.2. The tissue bath (section 7.4) is mounted onto an aluminum X-Y-positioning table (EK 100×80, Märzhäuser Wetzlar GmbH & Co. KG, Wetzlar, Germany). On top of the positioning table steal sheets are attached as a mounting base for magnetic posts. A sensor for measurement of extracellular potentials (section 7.3) can be positioned by a computer controlled micro manipulator (Patchstart, Scientifica Ltd., Uckfield, UK).

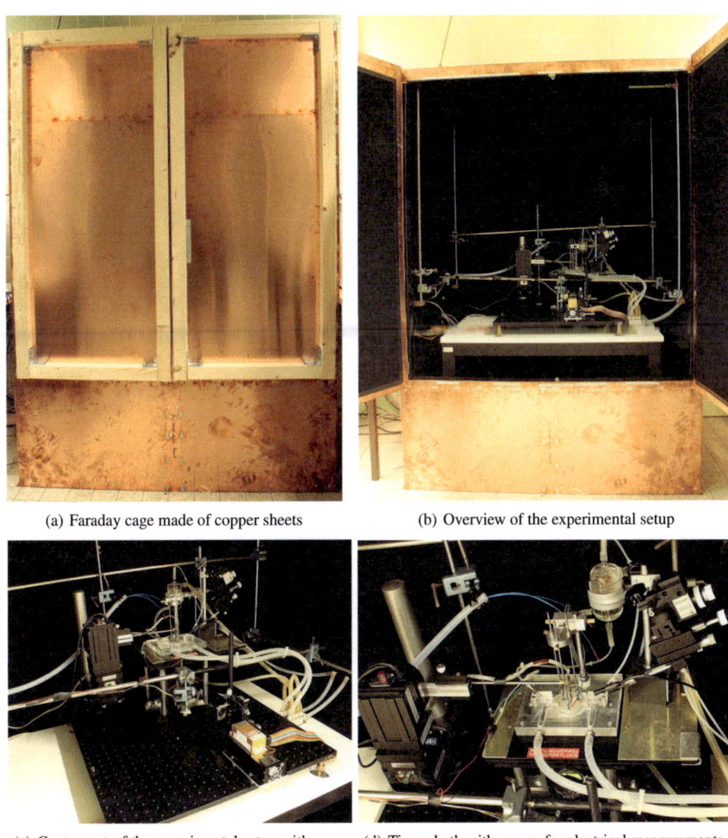

(a) Faraday cage made of copper sheets

(b) Overview of the experimental setup

(c) Center part of the experimental setup with tissue bath, micromanipulator, laser source and tube system for temperature regulation

(d) Tissue bath with sensor for electrical measurements, fiberoptic temperature sensor, glass heat exchanger and micro camera for electrode positioning

Figure 7.1. Photographic images of the experimental setup. The complete setup is enclosed by a Faraday cage for electromagnetic shielding. The inside of the cage is covered by light absorbing molleton. Detailed descriptions of the parts are given in the following sections.

7.1.2 Control and Automation

During a measurement sequence, several components need to be triggered in an automated way. In order to guarantee real time conditions within the sequence, the timing software was implemented on a microcontroller (atxmega 128a1, Atmel, San Jose, USA). Fig. 7.3 gives an overview of functional de-

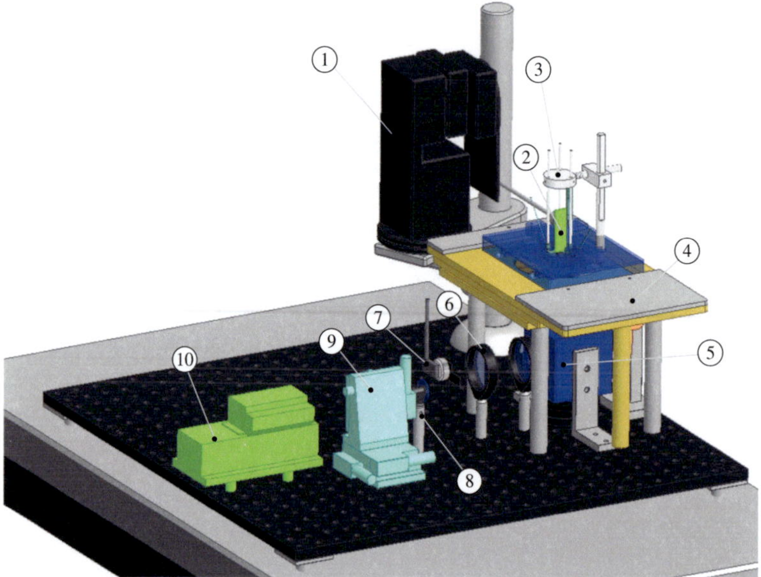

Figure 7.2. Overview of central part of the experimental setup. (1) Computer controlled micromanipulator for positioning of the experimental electrode; (2) Electrical measurement sensor with first amplifier stage; (3) Superfusion bath with mounting stage for tissue holder; (4) X-Y-positioning table with attached steal covers serving as platform for magnetic post mounts; (5) Filter cube containing a wavelength selective mirror, CCD camera is attached to this cube from underneath the table; (6) Lenses for laser beam forming; (7) Mechanical shutter to control illumination time, fixed to a mechanically uncoupled post (not shown here); (8) Circular neutral density filter for laser intensity adaptation; (9) Pinhole aperture for laser beam forming; (10) Laser source (green, 532 nm)

pendencies. Using a Matlab GUI [119] developed for this experimental setup (Fig. 7.4), acquisition parameters can be defined and recording started. The GUI takes as input parameters the stimulus frequency, maximum frame rate of the camera, interpolated frame rate, total recording time per sequence and the microcontroller frequency. Based on these values register values for the microcontroller are calculated and then transferred to the microcontroller via an RS232-interface. When acquisition is started, the GUI automatically creates a new folder for the acquired data and adds a log file, which contains all microcontroller settings.

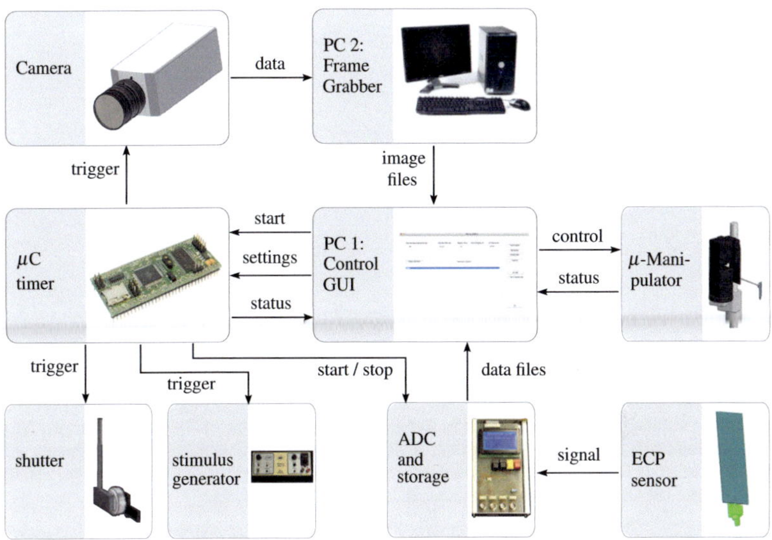

Figure 7.3. Overview of functional dependencies in the experimental setup. Register values for the microcontroller are calculated by a GUI on PC1. When a recording sequence is started, the microcontroller provides trigger signals for the camera, the mechanical shutter and the stimulus generator. Furthermore, it starts and stops the electrical data acquisition. Image data and electrical measurement data are transferred to PC1 after the measurement sequence. Electrode position can be varied remotely by sending control commands to the micromanipulator.

The microcontroller provides trigger signals for the mechanical shutter, which limits illumination time, the stimulus generator and the camera as well as start-stop information for the electrical measurement. A timing diagram for an exemplary recording sequence is given in Fig. 7.5. A master clock, which is determined by the set stimulation frequency initiates the recording sequence. Following the master clock the first image is acquired without illumination of the preparation. The shutter opens 5 ms before the second image, followed by three images with illumination. Coinciding with the 5th image, the preparation is stimulated by a unipolar stimulation pulse. The number of images is determined by the preset total acquisition time. The shutter closes 5 ms after the last image. One complete recording consists of a number of the described sequences, which provide the basis for the interpolated optical frame rate. Following the interpolation algorithm described in section 5.1.3

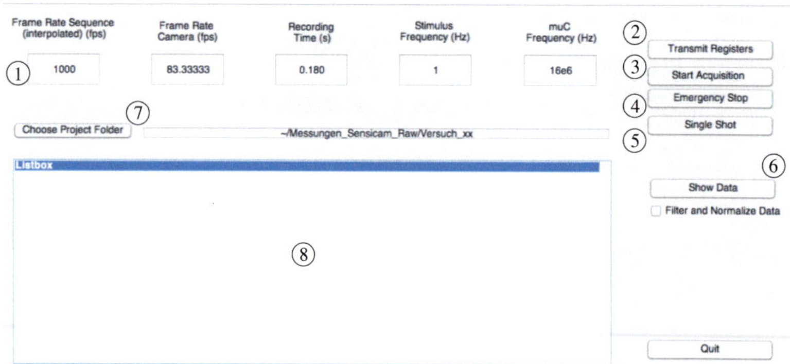

Figure 7.4. Matlab based control GUI for the experimental setup. (1) Input parameters for the recording sequence; (2) Button to start transmission of register values to the microcontroller; (3) Acquisition start; (4) Emergency stop button; (5) Acquire a single, laser illuminated image; (6) Show image data for the current recording in a separate GUI. (7) Destination folder for the current experiment; (8) Display for microcontroller status messages

the image acquisition is shifted relative to the stimulus pulse for each sequence. Consequently, the shutter opening is shifted in the same fashion.

For simultaneous measurement of extracellular potentials, the electrical data acquisition is started with the first master clock and stopped after the last image of the last recording sequence. When recording is completed, the microcontroller returns into an 'idle mode', in which it waits for new input data. In this mode, trigger pulses for the stimulus generator are continuously sent at the preset stimulation frequency.

Subsequently, the image files as well as the electrical measurement data can be transferred to PC 1, where a first analysis is performed (see section 7.2.2 and section 7.3.2).

7.2 Optical Measurement Part

Imaging of transmembrane voltages using voltage sensitive dyes (VSD) requires a low magnification optical system with optimized optical capabilities (see section 5.1.2). Briefly, a complete system should include an excitation

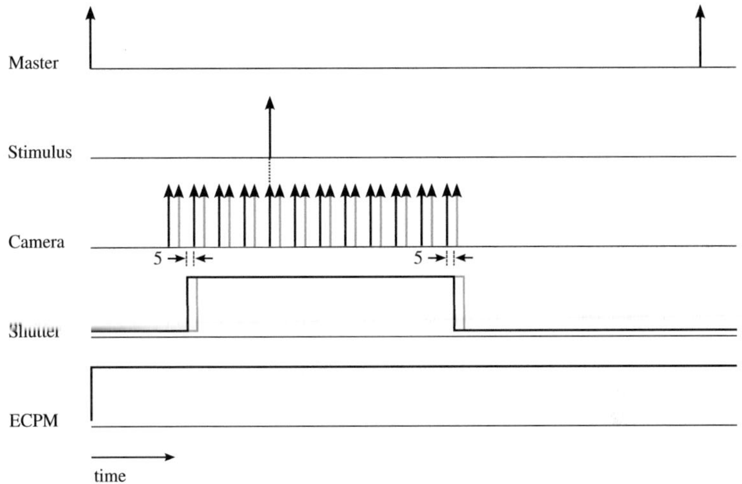

Figure 7.5. Timing diagram of one exemplary recording sequence. Shown numbers indicate times in ms. The master clock cycle depends on the set stimulation frequency. Following the master clock, trigger pulses for the camera, stimulus generator and the mechanical shutter are produced. Camera and shutter timing is shifted for each sequence to achieve an interpolated frame rate (grey lines). Upon the first master clock cycle of a recording the measurement of extracellular potentials (ECPM) is started.

light source, an optical system including high numerical aperture components as well as optical filters fitting the needs of the used VSD. For recordings a highly sensitive camera is needed.

In the following the optical part of the experimental setup is introduced. First, the optical components are described. Second, methods and software for the data acquisition are presented followed by a Graphical User Interface (GUI) for image processing and visualization.

7.2.1 Optical Components

Based on the works by Ratzlaff and Grinwald [85] and Castellana [120] the optical setup was built around a tandem lens macroscope formed by two video lenses with high numerical aperture. A complete overview of the optical part of the setup is depicted in Fig. 7.6. As excitation light source, a

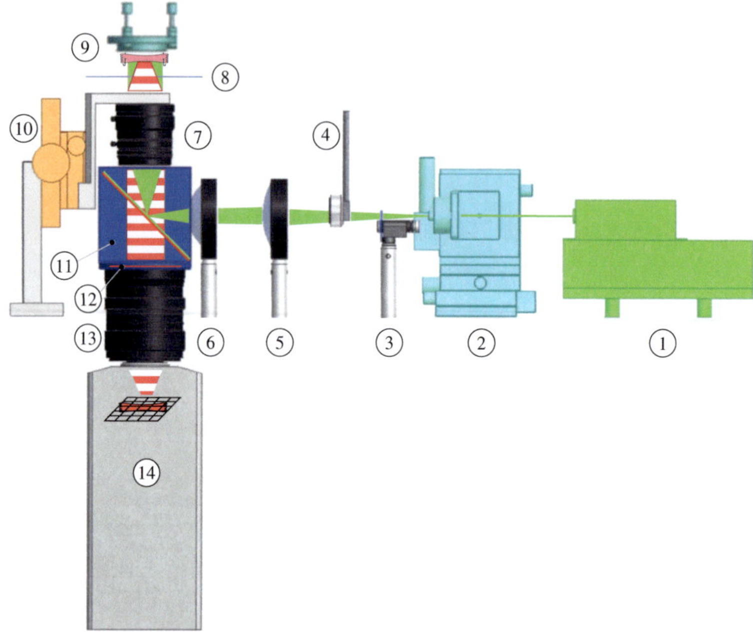

Figure 7.6. Overview of optical components in the measurement setup. Excitation light is produced by a green solid state laser (1) and coupled into the light path of a tandem lens macroscope. Emitted red fluorescent light is projected onto the chip of the CCD camera (14). Further components: (2) Pinhole aperture; (3) Adjustable neutral density filter; (4) Mechanical shutter; (5) Lens, f=100 mm; (6) Lens, f=60 mm; (7) Video lens, f-stop=1.4; (8) Glass slide (bottom of the tissue bath); (9) Tissue holder with myocardial preparation pinned to the silicon disk; (10) Linear positioning stage for focus adjustment; (11) Filter cube with a dichroic mirror, which reflects the laser light into the light path; (12) Longpass filter (610 nm) inside a custom made adaptor, which is attached to the filter cube; (13) Video lens, f-stop=0.95

green, continuous wave solid state laser (Compass 315M, Coherent, Inc. Santa Clara, CA, USA, wavelength: 532 nm, maximum output: 150 mW) is used. Its short term intensity fluctuations are smaller than 0.25 % of the preset intensity value. Following the light path trough the system, the laser beam is first expanded by a pinhole aperture. Fine adaptation of intensity can be done by a variable circular neutral density filter (Edmund Optics, Barrington, NJ, USA, diameter: 25mm, optical density: 0.04-1.5). Illumination time is controlled by a mechanical shutter, which can be controlled by a 15 V DC current. Afterwards, the laser beam is collimated by two plane-convex lenses (Thorlabs Inc.,Newton, NJ, USA, diameter 2", focal lengths: 60 mm, 100 mm). The beam is reflected by a dichroic mirror (Omega Optical, Brattleboro, VT, USA, LC576DRLP) before it reaches the focal point of a 50 mm video lens (NMV-50M1, Navitar Inc., Rochester, NY, USA, f-stop: 1.4). Emerging from the focal point the laser beam is being focussed to infinity by the lens and led onto the myocardial preparation inside the tissue bath, where it excites the fluorescent dye. The bottom of the bath is formed by a 1 mm glass slide (Carl Roth GmbH, Karlsruhe, Germany 76x52 mm).

The emitted fluorescent light is collected by the before mentioned video lens and focused to infinity. Due to its longer wavelength the emitted light can pass the dichroic mirror and reaches another longpass filter with a cut-on wavelength of 610 nm (Omega Optical, 635DF55). The longpass filter is mounted to a second 50 mm video lens (Navitar, DO-5095, f-stop: 0.95), which focusses the light onto the chip of a sensitive CCD-camera (Sensi-cam QE, PCO AG, Kelheim, Germany). A prerequisite for the choice of the video lenses was, that the first lenses' exit aperture (40.5 mm) was smaller than the entrance aperture of the second lens (50.5 mm), which avoids loss of brightness.

Optical filters were optimized for the specific needs of the VSD di-4-ANEPPS. The dichroic mirror was chosen with a high reflectivity at 532 nm and low reflectivity at wavelengths above 600 nm. An additional longpass filter was added to achieve a steeper cut-on edge at 610 nm.

7.2.2 Optical Data Analysis

Recorded image sequences of the VSD signal are analyzed in two steps. During the experiment a simple analysis is applied to check the quality of data. After the experiment further analysis is performed to extract local activation times and estimate the conduction velocity.

Analysis and Visualization During Experiments

For visual inspection of the VSD data during experiments, a simplified image processing workflow was implemented and included in the control GUI (Fig. 7.4). After the recording data is transferred to PC1. Image sequences are reordered according to the scheme explained in section 5.1.3. Subsequently, a Gaussian spatial filter and temporal filtering using a median filter are applied. The initial analysis is finalized by self-ratioing of pixel intensities (see section 5.1.4 for details). The processed data is visualized in a second custom developed Matlab GUI (Fig. 7.7). This GUI provides visualization for 3D data (2D plus time). Image data can be studied on a frame-by-frame basis. Additionally, the intensity of a selected pixel over time is plotted.

Further Data Analysis Steps

For a more detailed analysis, the data is filtered spatially by a Gaussian filter (10×10, filter mask with a standard deviation of 3). Afterwards, a bleaching correction is performed by subtracting the maximum for each recorded image sequence. Thereby, the maximum of each sequence is set to zero and therefore keeps this value during the scaling step. Subsequently, the sequences are reordered and a temporal filter is applied (median, window length: 5). Based on this data, a region of interest (ROI) can be selected by a manual selection tool. Within the selected ROI the local activation times are detected by the point of maximum downstroke velocity of the intensity signal (Fig. 11.3(b)). Based on the local activation times (LAT) a conduction velocity estimation is performed. In order to reduce outliers a 2D median filter of size 5 is applied onto the LATs (Fig. 7.8(d)). For a robust estimation of the CV first the barycenter for the largest connected LAT area belonging to one LAT is calculated. Based on the distance of the barycenters conduction velocities can be calculated.

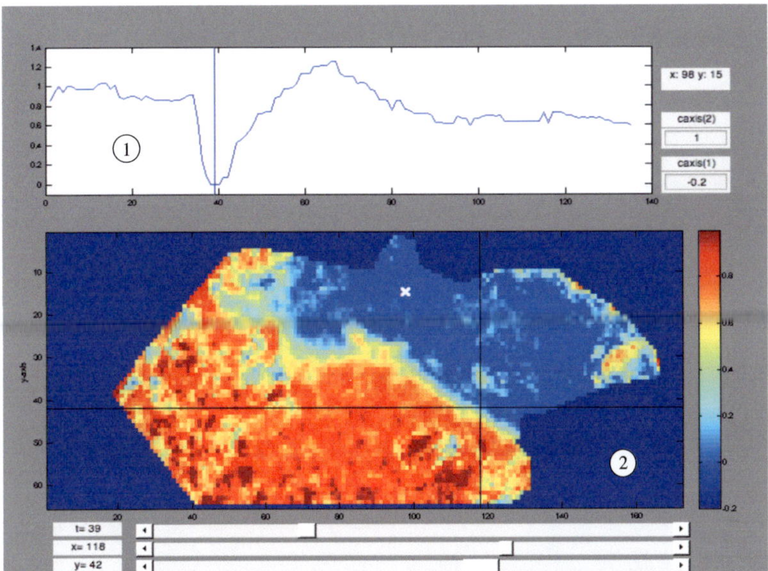

Figure 7.7. GUI for optical mapping data visualization. (1) Intensity course of one selected pixel over time. Time course is negatively correlated with the transmembrane voltage. Vertical line indicates the time point of the image frame shown in (2). (2) Visualization of one image frame; red colors show high intensity, corresponding to myocardium in rest. Light blue color indicates depolarized areas. Dark blue: Area outside the region of interest. White cross indicates the location of the selected pixel displayed in (1).

7.3 Electrical Measurement of Extracellular Potentials

In clinical practice, the only reliable method for functional diagnosis of cardiac electrophysiology is the measurement of extracellular potentials. The electrical measurement setup in the presented experiment was designed to reproduce the clinical approach as close as possible on a small scale level. An electrical measurement sensor was developed, which reproduces the geometrical constraints of a clinical catheter in orthogonal orientation relative to the tissue surface.

In this sensor, two platinum wires are used as measurement electrodes, which are directly connected to a first analogue amplifier stage. The amplified

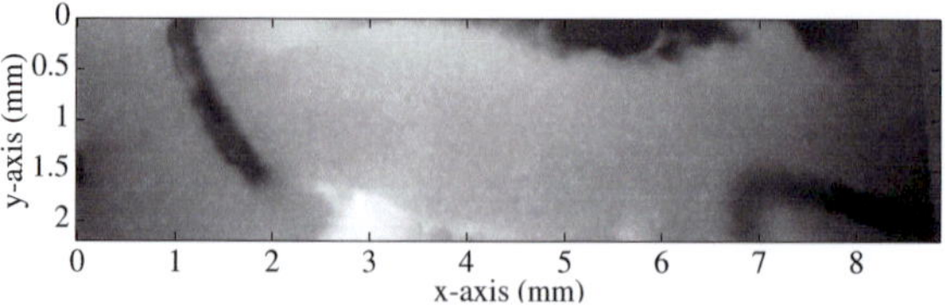

(a) Transmitted light image of a left ventricular papillary muscle. The muscle is fixed onto a silicone disc by tungsten needles (black).

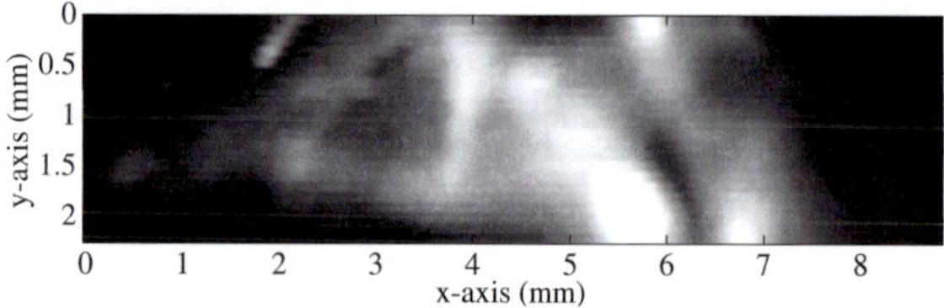

(b) Fluorescence signal emitted by the voltage sensitive dye

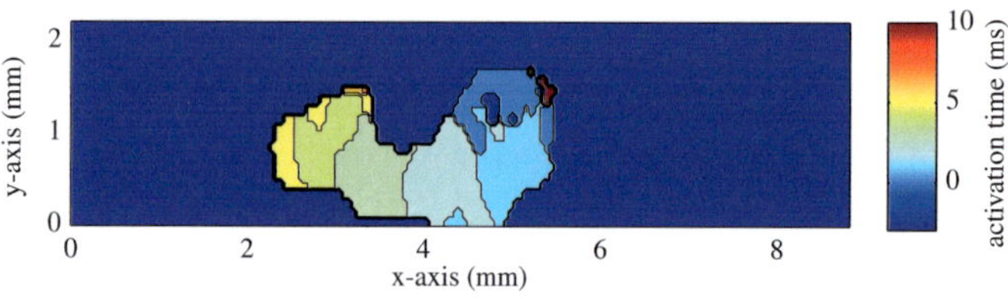

(c) Local activation times extracted from the fluorescence signal

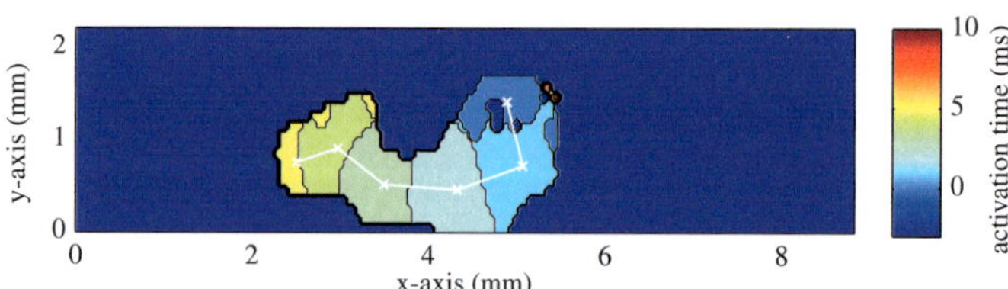

(d) Path along which CV is estimated on median filtered local activation times

Figure 7.8. Analysis of fluorescence data depicted for an exemplary left ventricular papillary muscle

signal is converted to digital values and stored by a custom made micro-controller board (Fig. 7.9(a)). The sensor is attached to a computer controlled

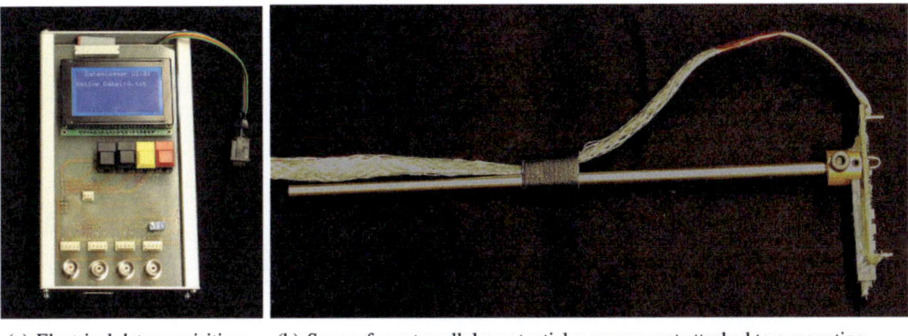

(a) Electrical data acquisition device

(b) Sensor for extracellular potential measurement attached to a mounting base

Figure 7.9. (a) Custom made electrical measurement device for AD-conversion of extracellular signals. (b) Mounted sensor for extracellular signal measurement

micro manipulator (Patchstar, Scientifica Ltd., Uckfield, UK) for precise positioning in three dimensions (Fig. 7.9(b), Fig. 7.2).

7.3.1 Microelectrode Design

A holder made from PVC (polyvinyl chloride) serves as a mounting base for the developed microelectrode sensor (Fig. 7.10). At the tip the holder contains two holes with an inner diameter of 0.45 mm, through which the wire electrodes can be fed. Measurement electrodes are made of a 0.35 mm diameter platinum wire. Platinum was chosen as electrode material, as it is the most frequent used material for ablation catheters in todays clinical practice (e.g. [121]). In many cases it is alloyed with iridium for improved hardness.

In order to get a smooth and even tip surface, the wire was fixed in between two sheets of PVC, containing a drilled hole and then cut using a fresh scalpel blade. The tip was gently deburred using a No. 1000 abrasive paper. The tip electrode was given a length of 2 mm compared to 1 mm for the reference electrode. This configuration was chosen, because clinical ablation catheters usually provide a longer tip electrode and shorter reference electrodes. The interelectrode spacing was set to 2 mm. In order to achieve a reliable electrical isolation, the drill holes were filled by cyanoacrylate based

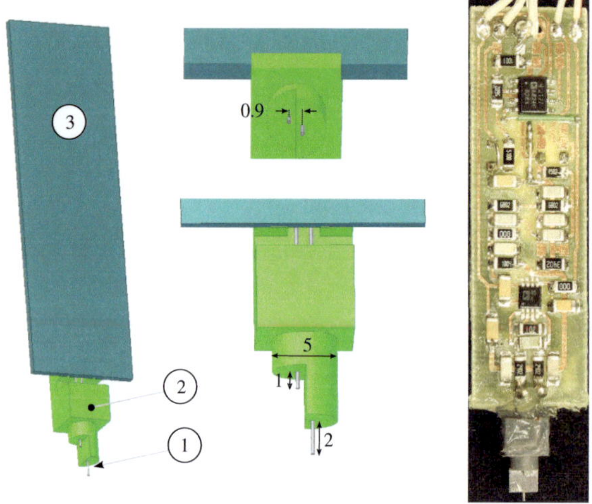

(a) Schematic visualization of the electrical measurement sensor   (b) Electrical sensor

Figure 7.10. (a) Different views of the electrical measurement sensor. (1) Platinum measurement electrode; (2) PVC based holder; (3) Electrical circuit board; (b) Bipolar version of the sensor with attached amplifier stage

glue, which was sucked into the holes by capillary attraction. The sensor was made waterproof by a clear lacquer coating. Additionally, the tip was enclosed in silicone.

Both platinum wires were welded to a circuit board containing an instrumentation amplifier and a first analog amplifier stage. The overall amplification was adjusted to a value of 625. An optional analog notch filter at 50 Hz turned out to be unnecessary to achieve good signal quality.

7.3.2 Data Acquisition

The sensor allows for a single channel differential measurement. On the one hand, unipolar signals between the distal electrode and a distant reference electrode can be measured. On the other hand bipolar electrograms from both tip electrodes can be received. Analog signals can be monitored online using a digital oscilloscope (Hameg HMO 1524, HAMEG Instruments

GmbH, Mainhausen, Germany). For AD-conversion initially a custom made device based on an 8-bit microcontroller (AVR ATxmega 128a1) was used. Thereby a sampling rate of 5 kHz was achieved, which is 4-5 times higher than in clinical recording systems. Alternatively, a National Instruments data acquisition system (NI-9205, National Instruments, Austin, USA) enabled sampling at a rate of 100 kHz. Digitized data was visualized and processed in a Matlab [119] based software.

7.4 Tissue Bath and Temperature Control

The experimental approach of simultaneous optical and electrical acquisition of electrophysiologic parameters of a myocardial preparation has require-ments, which were included in the design of a new tissue bath. The major constraints were:

- Optical observation of the preparation from underneath
- Accessibility of the preparation from above for electrical measurements
- Adjustable and temporally stable temperature inside the bath
- Easy mounting of the preparation
- Capability to mount various tissue geometries from 1D to 2D

These requirements were met by an optimized tissue bath made of acrylic glass, in which temperature is regulated by heated water, supplied by a tem-perature controlled circulating bath. Nutrient solution is provided through tubing and transported by a peristaltic pump. Before the nutrient solution enters the bath it passes a glass heat exchanger, which is also pre-warmed by the temperature controlled circulating bath. Bath temperature directly at the preparation is measured by a fiberoptic temperature sensor (Fig. 7.11 and Fig. 7.1(d)).

7.4.1 Tissue Bath Design

Various views and aspects of the tissue bath are depicted in Fig. 7.11. The body of the bath is formed by parts of acrylic glass. Two compartments were produced by milling. The first compartment is a circular shaped hole in the center, where the holder for the tissue preparation is positioned. This compartment is closed at the bottom with a 1 mm thick glass slide (Carl Roth GmbH, Karlsruhe, Germany 76x52 mm), which was fixed by silicone and allows observation from underneath. The outer walls of this compartment are in touch with the second compartment, which is connected to the temperature controlled circulating bath by silicone tubing. When constant flow of heated water through the outer compartment is established, the temperature in the inner compartment is stabilized. Inside the inner compartment the tissue holder is positioned by three steel rods, which provide screw threads to mount the holder (Fig. 7.11(a)).

The tissue holder was designed as a mounting base for various tissue geometries. It consists of a 4 mm thick silicone disk with a diameter of 3.5 cm, which is stabilized by an acrylic glass ring (Fig. 7.12). Depending on the tissue geometry and the way of access from above, different aperture geometries can be included in the silicone disk. Attached to the holder, three stainless steel screws are used to mount the holder inside the tissue bath. Due to the flexibility of the silicone and slightly oversized through-holes in the acrylic glass ring, the screws can be moved in a limited way. This feature allows adjustment of the plane, formed by the surface of the silicone disk, in a limited range. For mounting of the tissue preparation, the holder is placed upside down into a glass jar containing nutrient solution. In there, the preparation is pinned onto the silicone disk, using u-shaped tungsten needles. After mounting, the holder is placed into the tissue bath and temporarily kept in place by two acrylic glass sheets (see Fig. 7.11(b)). Subsequently, the steel rods can be screwed to the tissue holder and the acrylic glass sheets can be removed. Inlet and outlet for the nutrient solution is provided by two stainless steel tubes at both sides of the inner compartment. Filling level of the inner compartment can be adjusted by varying the height of the outlet tube.

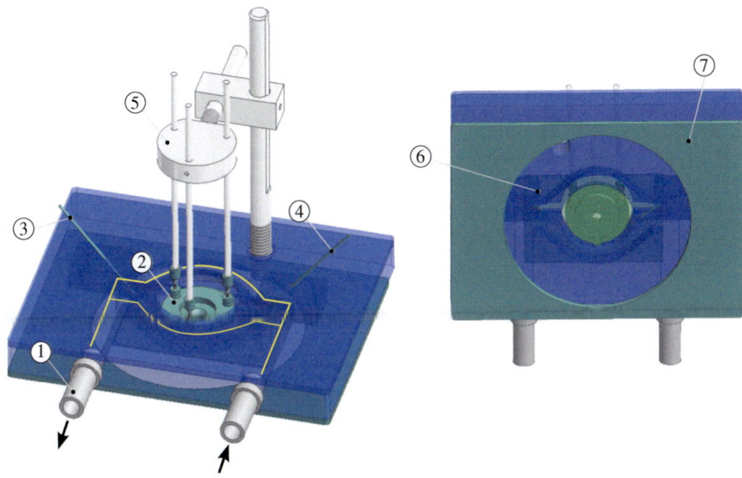

(a) Components of the tissue bath: (1) Inlet for heated solution, which is pumped through the included heat exchanging channels of the bath (yellow lines); (2) Silicone based tissue holder; (3) Inlet for nutrient solution; (4) Outflow for nutrient solution. (5) Mounting base for the tissue holder; (6) Glass slide for optical measurements; (7) Mounting frame fitting to the x-y-positioning table

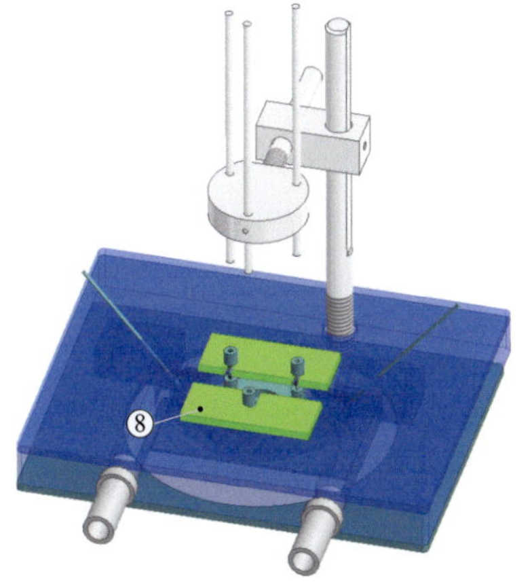

(b) Mounting of the holder inside the tissue bath. Holder with specimen can be rapidly transferred and fixed in an intermediate position by two holders (8, green). Subsequently, mounting rods can be screwed in.

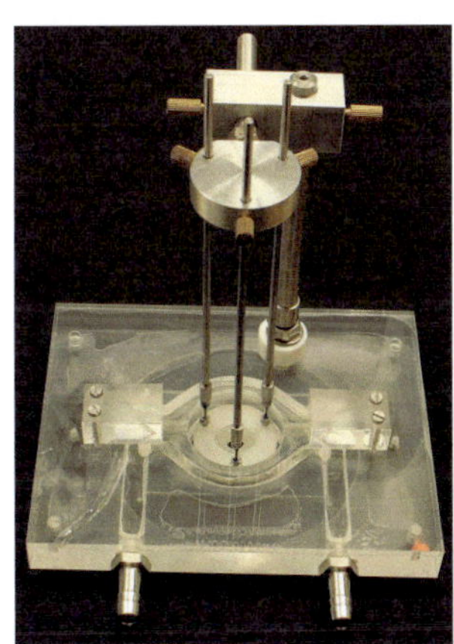

(c) Photographic image of the tissue bath

Figure 7.11. Schematics of the developed tissue bath for simultaneous optical and electrical measurements on cardiac preparations

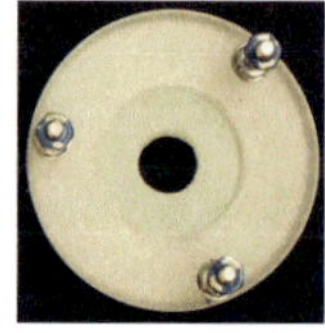
(a) Top view

(b) Bottom, side view

Figure 7.12. Tissue holder with circular aperture. Shown tissue holder is used for atrial preparations. For papillary muscles rectangular shaped apertures are used.

7.4.2 Temperature Inside the Tissue Bath

Keeping the temperature of the small amount of fluid absolutely stable over time is a demanding task. The problem was made easier by controlling the approximately two liters of distilled water inside a temperature controlled circulating bath (DC 10, Thermo Haake, Karlsruhe, Germany). Inside the circulating bath the temperature is kept stable in a range of 0.02 K. The distilled water is continuously circulated through a glass heat exchanger and the outer compartment of the tissue bath. After a stabilization time the temperature inside the inner compartment of the tissue bath is temporally stable within a range of 0.1 K. For all measurements an additional fiberoptic temperature probe (Fotemp, Optocon AG, Dresden, Germany) is placed adjacent to the tissue preparation to monitor the exact temperature.

7.5 Experimental Protocols

In this section protocols used in different parts of the experiments are described. The order was based on their temporal order during the experiment.

7.5.1 Preparation of Myocardial Specimen

Myocardial preparations used in this work were predominantly rat papillary muscles of the left and right ventricle. Additionally, in some experiments a preparation of the right atrium was established in a collaboration with the Institute of Biophysics at the Medical University of Graz.

A characterization of the region of interest is given in [96, 97]. All experiments were approved by the local committee for animal welfare.

Adult male Fisher Rats (F344) weighting 400-470 g were anesthetized by intraperitoneal injection of Ketamin (0.156 mg/g body weight) and Xylazin (0.014 mg/g body weight). After reaching deep anesthesia, hearts where quickly excised and placed into cooled Krebs-Henseleit solution (KHS). Subsequently the aorta was cannulated and the heart was perfused retrogradely for 10 minutes by cooled Krebs-Henseleit solution. Under continuos perfusion, papillary muscles and atrial preparations were dissected. One specimen was mounted to the tissue holder, the remaining were stored in cooled and continuously oxygenated solution for later use.

7.5.2 Krebs-Henseleit Solution

In all experiments Krebs-Henseleit (KHS) solution of the following composition was used (values in mmol/l): $NaCl$ 118.1, $C_6H_{12}O_6$ 11.1, KCl 4.7, $NaHCO_3$ 25.0, K_2HPO_4 1.2, $MgSO_4$ 1.2, $CaCl_2$ 1.8. During experiments solution was continuously gassed and oxygenated by Carbogen gas (95 % O_2, 5 % CO_2) in order to maintain a pH-value of 7.4±0.05. Monitoring of pH was continuos throughout the experiment. A concentration of 30 mmol/l of the uncoupling agent BDM (2,3-Butanedione-Monoxime, Carl Roth GmbH) was added to the storage and perfusion solution to reduce metabolic activity.

7.5.3 Stimulation

Stimulation was applied by tungsten electrodes in a unipolar fashion. A rectangular shaped current pulse of a duration of 1 ms was produced by constant current stimulus isolator (A 365, World Precision Instruments, Sarasota, Florida, USA). Stimulus amplitude can be varied between 5 μA and 10 mA. Stimulus amplitude was set to 1.5 times stimulation threshold.

7.5.4 Dye Loading

The voltage sensitive dye (di-4-ANEPPS, Sigma Aldrich, St. Louis, Missouri, USA) has to be dissolved in KHS for loading the preparation. The dye is shipped as powder in an amount of 5 mg. Upon arrival it was diluted in 1ml of DMSO (Dimethyl Sulfoxide, Sigma Aldrich) and subsequently divided and stored in aliquots of $25\,\mu l$. Storage of the aliquots at -21 °C was possible without alterations of the imaging potency of the dye for several months. During experiments one aliquot of dye was added to 25 ml of KHS. Preparations were superfused by the loading solution for 5 min followed by a washout of 10 min.

7.5.5 Acquisition of Temperature Profiles

During studies of temperature dependent electrophysiological changes, the temperature inside the bath was altered. Starting from a baseline temperature of 36.7 °C, the temperature of the circulating bath was increased in steps of 2 K. Thereby, inside the bath the following temperature steps were obtained: 36.7 °C, 38.4 °C, 40.3 °C, 42.0 °C, 43.8 °C, 36.7 °C. Temperature development was monitored using a fiber optic probe, positioned next to the specimen. Depending on the measurement position and the size of the preparation, temperatures at each level were reached with a maximum deviation of ±0.2 °C. Depending on the temperature level, it took 3-5 min to reach the next temperature step. After reaching a stable target temperature, data acquisition was started. After the last measurement sequence, temperature was returned to 36.7 °C. The time to reduce bath temperature to baseline level was 10-12 min.

8

Processing and Analysis of Clinical Intracardiac Electrograms

In order to validate the simulation with clinical human data, IEGMs were recorded during catheter ablation of typical right atrial flutter. During these procedures the cavotricuspidal isthmus is ablated to block the reentry pathway causing this arrhythmia. The signals were annotated by the physician during procedures and afterwards exported from the clinical recording system. Retrospective analysis was performed on physiological signals as well as signals recorded on RFA lesions of increasing size.

8.1 Clinical Data Acquisition

IEGMs from five patients were exported from the clinical recording system (EP Labsystem Pro, Bard, Natick, USA) at a sampling rate of 1 kHz. Each exported segment had a length of 60 s and contained catheter signals as well as surface ECGs . Catheter signals were acquired by an 8 polar coronary sinus catheter (CS) and an 8mm-tip 8F quadrupolar ablation catheter (Blazer Prime II, Boston Scientific, Boston, USA). For the latter, unipolar EGMs were exported additionally to bipolar EGMs. All signals were pre-filtered by a clinical standard filter setting, using a bandpass of 30-250 Hz. For three of the five datasets, also signals with reduced filtering applying only a highpass with a cutoff frequency of 1 Hz were exported. Catheter orientation was determined from fluoroscopic images. This annotation was verified by signal amplitudes and LATs of the unipolar EGMs during the analysis.

8.1.1 Electrograms Recorded During Ablation

Electrogram alterations during the RFA sequence were studied in five patients. Four of the five patients were paced from the CS electrodes 7 and 8, at a basic cycle length of 500 ms in order to ensure comparable excitation patterns. The fifth patient presented in sinus rhythm. Ablation was performed in a point-by-point strategy. EGMs were exported for point ablations where no overlap to previous lesions was present. Ablation duration varied between 20 s and 60 s. EGMs for each ablation point the following sequence of signals was evaluated: (1) 10-20 s before ablation, (2) 20-60 s during ablation and (3) 20 s post ablation. Only catheter positions which were stable during the complete ablation sequence were accepted for analysis. Overall 11 complete datasets were analyzed containing the following orientations.

- 6 sequences with non-parallel catheter orientation
- 2 sequences with parallel catheter orientation and propagation from the proximal to distal electrode (para-pd)
- 3 sequences with parallel catheter orientation and propagation from the distal to proximal electrode (para-dp)

8.1.2 Physiological Signals with Varying Catheter Positions

In one patient additional EGMs recorded in 13 different catheter orientations and varying electrode tissue distances were acquired on normal tissue to gain information on position related changes in the EGM. Different classes were distinguished for annotation:

- Parallel/ non-pararallel orientation
- Non-contact/ contact/ contact with pressure (indented)

Segments of 60 s duration were exported for verification of simulation results for each catheter orientation.

8.2 Workflow for Electrogram Segmentation and Template Generation

Clinical signals were used to generate templates, which provide a good basis for morphological analysis and comparison to simulated data. In a first step EGMs were extracted using an automatic segmentation algorithm based on the NLEO (see Sec. 4.3). In the case of pacing, also the paced lead (CS78) was included in the analysis to improve the robustness of detection. Based on the detected activations in the pacing signal (Fig. 8.1(a)), those bipolar EGMs were selected, which directly followed the stimulus. If no pacing was applied, active segments (AS) of the bipolar lead were taken in the first step of the segmentation. An exemplary segmentation is displayed in Fig. 8.1(c). Only electrograms, which agreed with the detected dominant frequency were included in the analysis. AS, which fit to the corresponding time lag of the dominant frequency were selected by an algorithm introduced in section 4.3.

Onsets of bipolar EGMs were used to extract EGMs for further analysis. As the NLEO based segmentation outlines the EGMs quite narrowly, a longer portion of the signal, starting 25 ms before the activation with a window length of 125 ms, was extracted (Fig. 8.1(b)). The same time frame was applied on the corresponding unipolar EGMs. Preceding the creation of templates, bipolar EGMs within 5 s segments were aligned by maximum correlation (Fig. 8.1(d)). Time dependency between all leads was preserved by applying the same alignment to the unipolar EGMs. A representative EGM template was obtained taking the mean of the aligned EGMs (Fig. 8.1(c)). Chosen 5 s windows did not overlap, which e.g. resulted in 4 templates obtained from a 20 s signal.

8.3 Estimation of the Clinical Filter Transfer Function

The exact knowledge of the filter function used in a recording system is necessary for a detailed comparison of simulated and clinical data. As this function is not provided by the manufacturers, a system identification was

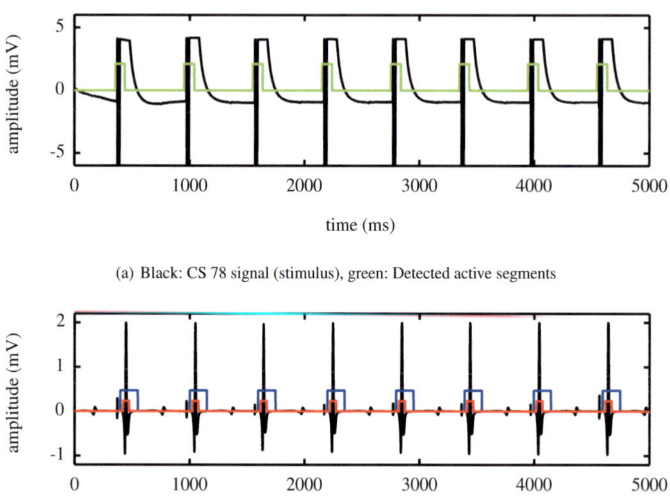

(a) Black: CS 78 signal (stimulus), green: Detected active segments

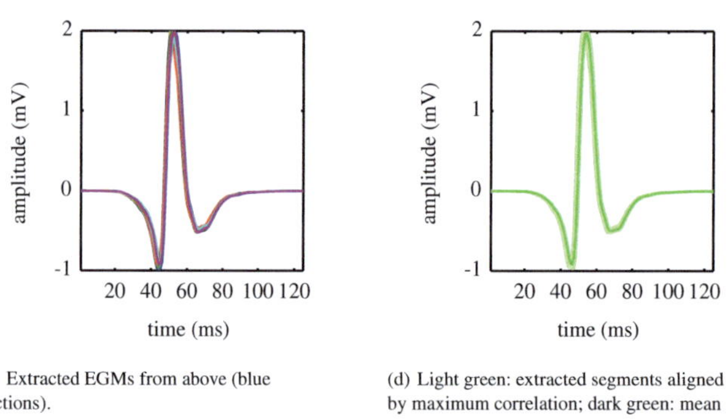

(b) Black: Bipolar EGM with active segments (red) and identified EGMs (blue). Blue segments were used for template generation.

(c) Extracted EGMs from above (blue sections).

(d) Light green: extracted segments aligned by maximum correlation; dark green: mean template of the active segments

Figure 8.1. Workflow for template generation over a signal segment of 5 s. (a) Stimulus signal from CS78 with segmentation. For each detected activation the next active segments in the bipolar signal (b), which fits to the dominant frequency was included in the analysis. Subsequently, identified signal pieces were extracted (c) and aligned by maximum correlation (d). The mean of the aligned segments was used as a representative template.

necessary. One approach is described in [122]: The system transfer function $H(f)$ can be determined from the power spectral density (PSD) of the input signal ($S_{xx}(f)$) and the cross PSD of output and input signal ($S_{yx}(f)$):

$$H(f) = \frac{S_{yx}(f)}{S_{xx}(f)} \tag{8.1}$$

$H(f)$ represents the filter function of the clinical recording system, where x(t) is an unfiltered signal and y(t) is a signal filtered by the specific system.

In order to determine the filter transfer function one signal with a length of 20 s was analyzed in two versions. The first one with standard filter settings (30-250 Hz), the second one was only highpass filtered at 1 Hz to prevent it from exceeding the measurement range. Using these signals, the PSDs were calculated by the averaging method proposed by Welch [123]. Applying this method, a 2048 sample Hamming window was used to partition the signals with an overlap of 50 %. Phase and magnitude of the estimated transfer function were compared to several standard filters (Butterworth, Chebyshev, Elliptic) of different orders. The filter providing the best match was applied to 80 unfiltered electrograms for validation. The results were then correlated with their versions filtered by the clinical system.

9

Simulation of Intracardiac Electrograms

9.1 Simulation Setups

This section gives an overview of methods and parameters related to simulation setups. These parameters cover geometrical aspects and material parameters. First, an introduction of all simulated catheters and measurement electrodes is given. Second, different aspects on physiological and pathological substrate modeling are described.

9.1.1 Geometrical Models of Catheters and Measurement Electrodes

In this work, several geometries and sizes of electrodes measuring extracellular cardiac potentials have been applied in simulation. The fields of application of these electrodes range from experimental, to clinical and finally to virtual catheters simulated for developmental purposes.

Experimental Microelectrode

An exact model of the experimental measurement electrode introduced in section 7.3.1 was generated for simulation (Fig. 9.1). Electrode lengths and interelectrode spacing represent approximate values of clinical catheters. Electrode diameter was adjusted to match the size of preparation used in the experiment.

Catheters in Clinical Use

Catheters used in clinical procedures differ in size and shape depending on the specific application. For diagnostic purposes multi-electrode catheters

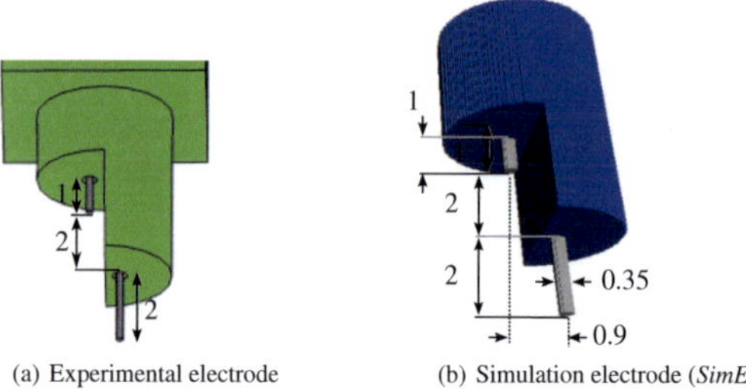

(a) Experimental electrode   (b) Simulation electrode (*SimEl*)

Figure 9.1. (a) CAD drawing of the measurement electrode used in the experimental setup. Basic framework made from plastics (green); Platinum wires with a diameter of 0.35 mm (gray) (b) Representation of the experimental electrode used in simulation with electrodes (gray) and isolating material (blue); numbers indicate lengths in mm

are used to provide several recording points at once. Catheter manufacturers offer a variety of multi-electrode catheters with different diameters, shapes as well as electrode widths and spacings. Signal shapes produced by this group of catheters were studied using the geometries shown in Fig. 9.2. Most common electrode widths (1 mm and 1.2 mm) and electrode spacings of 1, 2, 3 and 5 mm could be studied with these geometries. Electrodes of this type are used in circular or finger like catheters as well as in basket catheters (see section 9.1.1).

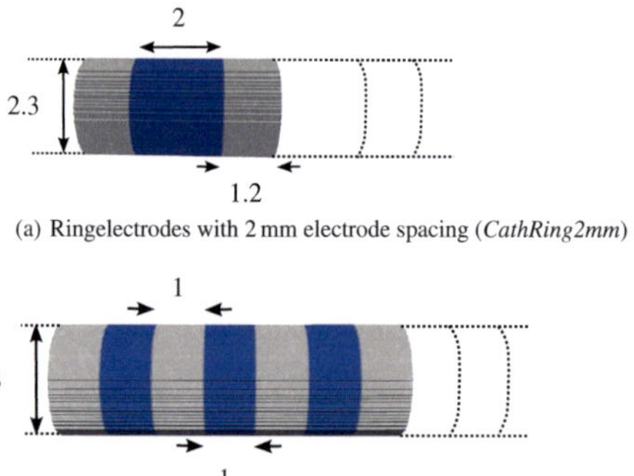

(a) Ringelectrodes with 2 mm electrode spacing (*CathRing2mm*)

(b) Several ringelectrodes with 1 mm spacing (*CathRing1mm*)

Figure 9.2. Models for clinical mapping catheter electrodes, as used in circular, finger-like and basket catheters. Electrodes in gray are separated by isolation (blue). Numbers indicate lengths in mm.

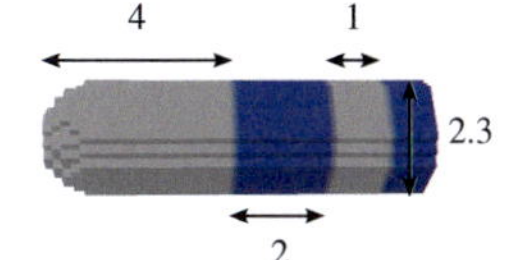

(a) Ablation catheter with a 4 mm tip electrode (*CathAblation4mm*)

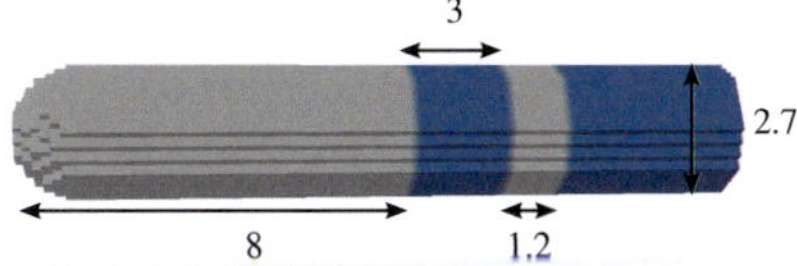

(b) Ablation catheter with an 8 mm tip electrode (*CathAblation8mm*)

Figure 9.3. Two models of ablation catheters. 4 mm tip electrodes are used for substrate based mapping. 8 mm electrodes create large lesions and are therefore suitable to create line shaped lesions. Numbers indicate lengths in mm.

For RF-ablation therapy, catheters with a longer tip electrode are used in order to achieve a larger contact area between electrode and tissue. Tip electrodes with lengths of 4 mm and 8 mm are used in clinical procedures. Two common ablation catheter designs were simulated in this work (Fig. 9.3).

New Catheter Designs

Today available catheters for clinical use still give room for improvements regarding new designs optimized for specific applications. Two new designs were investigated in the simulation environments developed in this work. The first one is a new ablation electrode design containing three microelectrodes (Fig. 9.4). This design was provided by Biosense Webster. Inside the surface of a common ablation electrode with a diameter of 2.3 mm three microelectrodes were placed in a triangular arrangement. Each microelectrode with a diameter of 0.18 mm is insulated against the ablation electrode. Analysis of the microelectrode signals is supposed to reveal spatial information, which is not visible in the signal of the 'large' ablation electrode due to spatial averaging. The three microelectrodes offer additional information on local tissue characteristics at the catheter tip. The development and optimization of this new catheter design was supported by a simulation study described later in this chapter.

A second catheter design was developed in this thesis for improved control of RF application during RFA. Different from regular ablation catheters the ablation electrode is not located at the catheter tip, but in the center of four ring electrodes (Fig. 9.5). Ring electrode signals are supposed to deliver improved information on lesion width and depth during the ablation process.

9.1.2 Simulation Parameters Under Physiological Conditions

IEGM for the introduced catheters were studied under physiological tissue conditions. For this purpose, geometric parameters like myocardial patch size and thickness as well as catheter orientation relative to the tissue surface were varied. Furthermore, conductivities for all used materials were defined. A schematic showing important geometric parameters is given in Fig. 9.6.

Tissue Dimensions

Tissue geometry for all simulations in this work was a 3D planar patch of myocardium with a width w (y-direction) and a length l (x-direction). These parameters were chosen according to the constraints of the respective simulation study. In the majority of simulations, propagation direction was along the x-axis of the setup. Therefore w was set in a manner, that for each regarded catheter orientation a space of at least 10 voxels was in between the

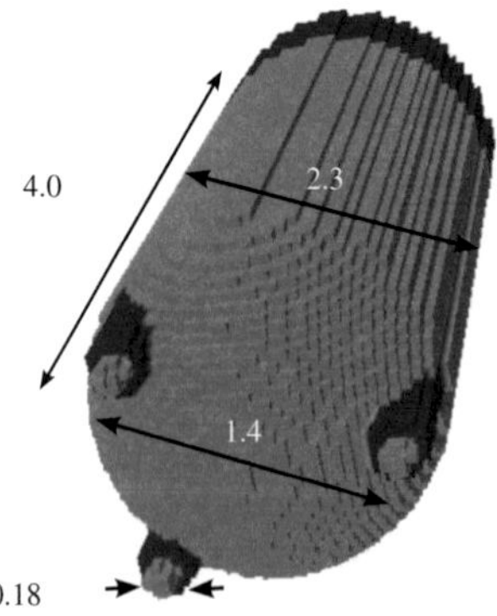

Figure 9.4. Ablation catheter with three integrated microelectrodes (*Micro-Catheter*)

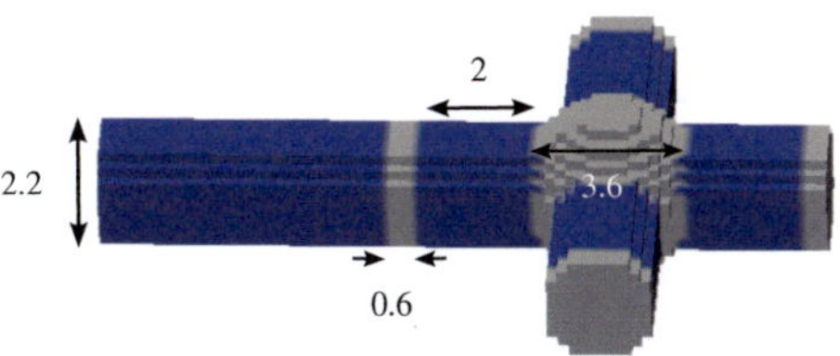

Figure 9.5. A potential design for an ablation catheter with one centered ablation electrode surrounded by four measurement electrodes (*Cross-Catheter*). This design for a catheter is supposed to provide four electrodes surrounding the lesion for improved EGM based lesion assessment.

electrode and the border of the patch. The choice of l was a tradeoff between the field of view of the catheter electrodes and computation time. The electric field of a source declines with the distance between source and point of interest. Therefore l was set up to 40 mm if simulation time was still acceptable.

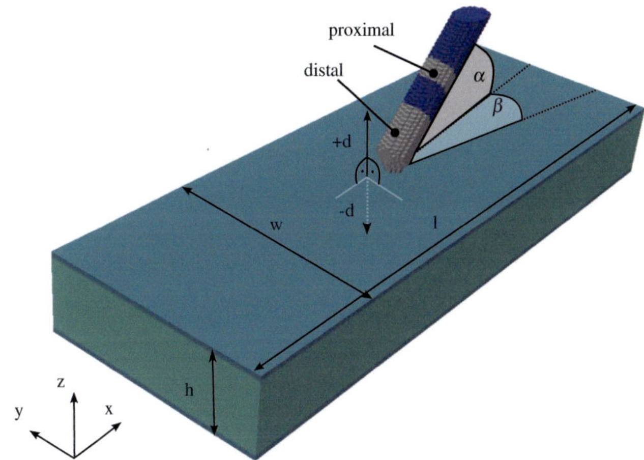

Figure 9.6. Schematic of a representative simulation setup. A catheter is positioned on a planar patch of myocardium (green), which is covered by a 2 voxel layer of connective tissue (dark green). w, l, h represent the tissue dimensions. The catheter (electrodes: gray, isolation: blue) is tilted in the x-z-plane relative to the x-axis by an angle α. The orientation in the x-y-plane relative to the x-axis is given by β. Tissue electrode distance is given by d, which is positive if the catheter is not in contact and negative if the catheter penetrates the myocardium.

The height h of the patch was set in the range of the myocardial wall thickness. In most cases atrial EGMs were studied. Therefore, it was set to the range of 2-3 mm, which was found as mean value of the human atrium [124–126]. The endo- and epicardial side of the patch were covered by a 2 voxel layer of passive connective tissue, which accounted for the properties of these layers [16].

Catheter Orientation and Position

Catheter orientation was defined relative to the tissue surface (see Fig. 9.6). It was set by the tilt angles projected onto x-y-plane (α) and x-z-plane (β) as well as the distance between catheter tip and tissue surface (d). Negative values of d imply a deformation of the tissue surface by the indented catheter tip.

Conductivity Values

Used conductivity values for the basic materials are given in Tab. 9.1. Extracellular values for blood and extracellular space of myocardium were set to literature values [127–129]. As endo- and epicardial surface layers were modeled passively without intracellular connection to heart cells, conductivity only needed to be defined for the extracellular space. This value was set equal to myocardium. Conductivities of electrodes e.g. made of platinum is usually above 10^6 S/m [130]. However numerical stability of the simulation program is not given for these high conductivities. Therefore, the electrode conductivity was set to the 10000 fold conductivity assumed for blood. At this conductivity the electrode still formed an equipotential volume for the regarded time steps. Conductivity of isolation was set to a value in the range of teflon [131], which is a common material used for clinical catheters.

Intracellular conductivity of myocardium represents the cytosolic and intercellular conductivities across the gap junctions (see section 2.1.1). This parameter was adjusted in order to match conduction velocities reported for human myocardium of 0.3-1.2 m/s [132]. For each simulation study it was defined to specific constraints and is therefore indicated as variable in this

Table 9.1. Conductivities used for simulation; extracellular conductivities σ_e, intracellular conductivities σ_i in S/m

	σ_i	σ_e
Myocardium	variable	0.2
Connective tissue	0	0.2
Blood	0	0.7
Electrode	0	$7 \cdot 10^4$
Isolation	0	10^{-22}

section. In order to apply no-flux boundary conditions at the borders for the myocardial patch, intracellular conductivity of surrounding materials was set to zero.

9.1.3 Modeling of Fibrotic Substrates

Aiming to study electrogram fractionation, myocardium with fibrotic patterns in 3D was produced. Two types of fibrosis were reproduced based on the definitions given by [25] (see section 2.2.2). Diffuse fibrosis was described as homogeneously distributed very small fibrotic areas. Patchy fibrosis is formed by strand like elements of up to several millimeters, which are mostly aligned along fiber direction. As suggested by Spach et al. [112] and Jacquemet et al. [28] for the 2D case, fibrotic elements were distributed and shaped following random distributions. On a regular spaced grid with a spatial resolution of $50\,\mu$m, seed points were created by uniform distributions of x, y and z coordinates in the myocardial volume. Each seed point was replaced by a fibrotic element. For diffuse fibrosis, these elements were cubes of $2\times2\times2$ voxels. Patchy fibrosis was created by square shaped disk like elements with a fixed thickness of 2 voxels. Length (y) and depth (z) of the disks were defined by Poisson distributions. The Poisson distributions were defined by the expectancy value λ . In y-direction λ_y was set to $900\,\mu$m as proposed in [28, 112]. In z-direction λ_z was reduced to $300\,\mu$m to account for the limited thickness of the myocardial wall.

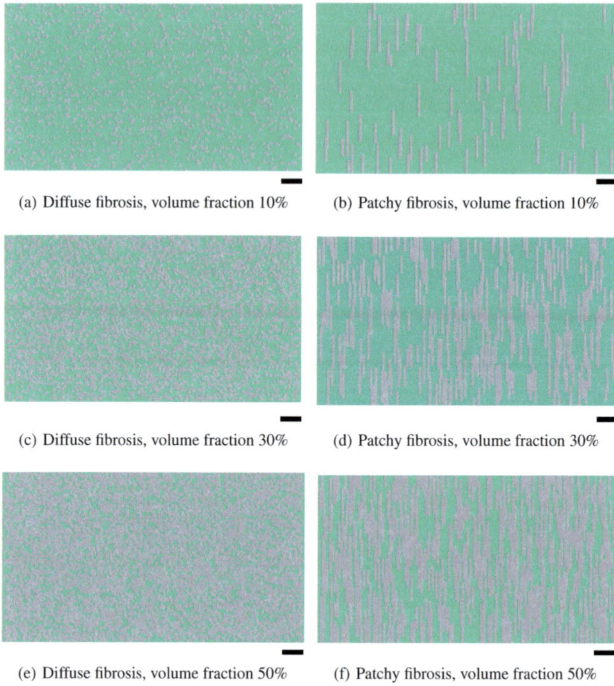

(a) Diffuse fibrosis, volume fraction 10%   (b) Patchy fibrosis, volume fraction 10%

(c) Diffuse fibrosis, volume fraction 30%   (d) Patchy fibrosis, volume fraction 30%

(e) Diffuse fibrosis, volume fraction 50%   (f) Patchy fibrosis, volume fraction 50%

Figure 9.7. X-Y-plane views of simulation setups for diffuse and patchy fibrosis of different volume fractions. Fibrotic areas are indicated in gray inside myocardium (green). Black bars represent a length of 1 mm

Fibrotic elements were modeled as passive areas with high intracellular resistance (10^{-10} S/m), extracellular conductivity was set to 10^{-3} S/m assuming, that fibrotic elements do not consist of pure collagen, which has a conductivity in the range of 10^{-18} [133]. In this model of fibrosis, only structural parameters were studied. Therefore, myocardial conductivity and sodium handling were not changed, as done in other studies [28, 29].

9.1.4 Modeling of Acute Ablation Lesions

Lesions created by RFA result from local heating due to a high electric current density at the catheter tip. Lethal temperatures leading to direct cell

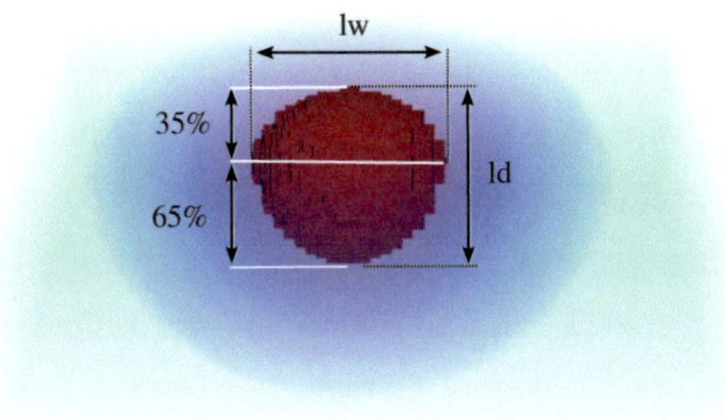

Figure 9.8. 3D geometry of an acute RFA lesion surrounded by myocardium (green). The inner lesion core (red) is formed by two half ellipsoids accounting for 35% and 65% of the lesion depth (*ld*). The lesion width (*lw*) is determined by a fixed width to depth ratio. The inner core is surrounded by vital tissue with changed electrophysiological properties (red to blue).

death are reported for temperatures above 50-60 °C [37, 49, 134]. For the acute lesion, the tissue surrounding this inner core is heated and therefore temporally or permanently changes its electrophysiological properties. This section describes an electrophysiological model for acute ablation lesions, which forms the basis for EGM studies in the lesion area.

Lesion Geometry

Literature data introduced in section 3 showed that lesion shape of RFA-lesions can be approximated by an ellipsoidal shape. In this work lesion geometry of the necrotic core consisted of two half ellipsoids set on top of each other (Fig. 9.8). The upper half ellipsoid, facing towards the endocardial surface accounted for 35% of the lesion depth (*ld*). The lower ellipsoid represented the remaining 65%. This shape closely resembles the extensive modeling and experimental data shown in [135]. While lesion depth was set in order to model scar formation over time, lesion width (*lw*) was calculated by a fixed width to depth ratio. As introduced in section 3 this ratio can vary between 1.25 and 1.9. In order to account for the borderzone of heated but still vital tissue, six layers representing different temperature zones (49-

40 °C), were defined surrounding the necrotic core (Figs. 9.8, 9.9). Based on the fact that temporal or permanent loss of excitability occurs for temperatures above 50 °C, the outer layer of the passive necrotic core was assumed to coincide with this temperature. Temperature profile from 50 °C to 40 °C was modeled as exponential decrease within a distance of 3 mm. This resembles the simulation data shown in [136] and [137]. Extracellular conductivities in the borderzone were set equal to that of myocardium to 0.2 S/m. For the lesion core, which contains denatured tissue an increase in resistance for low frequencies was reported [138]. Therefore, extracellular conductivity was set to 0.1 S/m.

Action Potential Changes

For temperatures above 37 °C, AP parameters gradually change (see section 3). Most pronounced changes are (1) increased resting membrane voltage (V_m), (2) reduced maximum upstroke velocity (dV/dt_{max}), (3) reduced AP amplitude (APA) and (4) shortened AP duration (APD). These changes are macroscopically reproduced by electrophysiological models adapted to represent ischemic tissue. Therefore, a version of the ten Tusscher ventricular electrophysiological model [104] adapted to ischemia phase 1a by Wilhelms et al. [105], was used (for details see section 6.1.2). In this model ischemic effects can be weighted by a zone factor. Depending on the zone factor the three ischemia effects hyperkalemia, acidosis and hypoxia, are considered in the cell model. For ischemia, these effects are weighted by a non linear curve depending on the position in the ischemic region. These curves were linearized to achieve a linear increase of AP changes, when approaching the lesion core. For the acute lesion model, AP changes were set to increase linearly for the temperature range of 43-50 °C (Fig. 9.9). AP characteristics for each temperature zone were first adjusted in simulations of tissue strands ($50 \times 10 \times 10$ voxels). However, these values are highly dependent on tissue geometry. Therefore, a further tuning of parameters was performed in a full lesion geometry.

Conduction Velocity Changes

For increasing tissue temperatures, CV first increases in the range of 38-42 °C up to 125% of baseline CV. Afterwards it starts to decrease reaching about 60% at 49 °C, before conduction is blocked at 50 °C [139]. A certain CV reduction results from the reduced upstroke velocity introduced in the last paragraph. A further target for CV adjustment is intracellular conductivity. The rationale behind this is the strong variability of gap junctional conductivity for increased temperatures [49]. In order to model such a CV profile in the lesion border zone, further tissue strand simulations were performed to obtain initial values of intracellular conductivities. These values were further adjusted in the full lesion geometry.

9.2 Simulation of Intracardiac Electrograms

In chapter 6, mathematical methods to simulate extracellular potentials and transmembrane voltages were introduced. Based on these parameters IEGM were calculated in this work. This section first introduces definitions for simulated unipolar and bipolar electrograms and then shows measures to analyze fiducial points in the extracted signals. Finally, an approach to calculate extracellular currents is described.

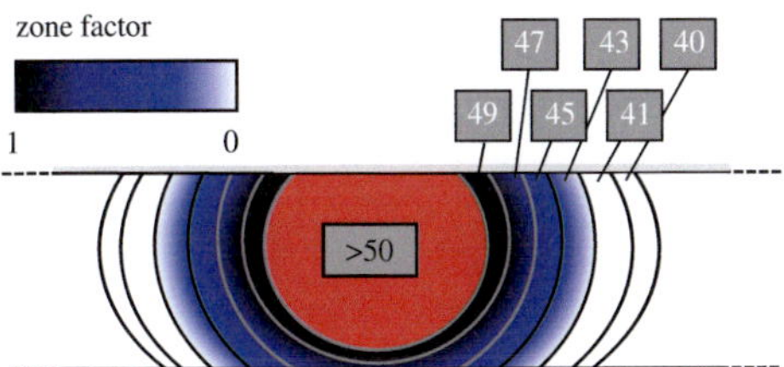

Figure 9.9. Schematic X-Z-plane view of a modeled acute ablation lesion. Numbers (40-50) indicate temperatures in °C. The passive necrotic core (red) is surrounded by six layers of heated tissue with changed conduction velocities. AP related effects are weighted by a zone factor and increase towards the lesion core.

9.2.1 Electrogram Definitions

Clinically, most intracardiac signals are measured bipolar (v_{bip}). In this case the potential of the distal electrode (p_{dist}) is subtracted from the proximal one (p_{prox}).

$$v_{bip}(t) = p_{prox}(t) - p_{dist}(t) \tag{9.1}$$

Simulated IEGM were obtained as well by subtracting the potentials extracted from the virtual catheter electrodes. In clinical practice, unipolar electrograms (v_{uni}) are measured between single electrode potentials (p_{el}) and the Wilson's Central Terminal (see section 4.1.1). For the simulated signals a reference (p_{ref}) was calculated by the mean of all potentials in the blood in the top most X-Y-slice of the setup.

$$v_{uni}(t) = p_{el}(t) - p_{ref}(t) \tag{9.2}$$

Simulated signals were stored at a sampling rate of 10 kHz, which is the 5-10 fold value used in today's clinical recording systems.

9.2.2 Simulation Data Analysis

Fiducial points in the signal were used to quantify and compare signal changes. By an automatic annotation the following points were delineated:

- Maximum positive peak (p_{pos})
- Maximum negative peak (p_{neg})
- Maximum absolute deflection (dV/dt_{max}) between (p_{pos}) and (p_{neg})

The described points were analyzed for raw signals and for simulated signals filtered with the clinical filter function described in section 8.3.

9.2.3 Calculation of Extracellular Currents

With the knowledge of the extracellular potentials (ϕ_{extra}) and the respective conductivities, extracellular current flow was studied. Using the Poisson Equation of electrostatics the electric field $\mathbf{E}_{extra}$ was calculated.

$$\mathbf{E}_{\text{extra}} = -\nabla \phi_{\text{extra}} \tag{9.3}$$

Subsequently the extracellular current density j_{extra} was calculated by:

$$\mathbf{j}_{\text{extra}} = \mathbf{E}_{\text{extra}} \cdot \sigma_{\text{e}} \tag{9.4}$$

Thereby, σ_{e} can be a scalar for isotropic conditions or a tensor in the case of anisotropy. Analogous, the intracellular current can be calculated from the intracellular potentials ϕ_{i}. An exemplary visualization of intra- and extracellular currents is shown in Fig. 4.1.

Part III

Results

10

Intracardiac Electrograms Recorded by Clinical Catheters Under Physiological Conditions

Interpreting and understanding the morphology of intracardiac EGMs is a key challenge for future advances in electrophysiological procedures. This chapter illustrates simulated EGMs under physiological conditions. It demonstrates the applicability of the chosen simulation framework to confirm and explain clinically measured EGM morphology. The first part covers a comparison of clinical and simulated EGMs for verification of the chosen approach. In the second part, this method was used to study EGM morphology subject to varying catheter orientation, tissue thickness and conduction velocity. Finally, the influence of different filter settings common in clinical practice was investigated.

10.1 Applied Methods

Evaluation of Dirichlet Boundary Conditions for the Simulation of Intracardiac EGM

It is common practice to define a Dirichlet boundary condition (see section 6.2.1) in the simulation of extracellular potentials, in order to define their absolute values. These are defined at one or several points in the simulated volume. It was found, that this boundary conditions in the small scale simulation setups used in this work can lead to electrical field distortions in the solution, which did affect EGM shape. Furthermore, in the used simulation software package [114], simulation times significantly increased, when a Dirichlet boundary condition was added. In the case, in which only no-flux

Neumann boundary conditions are applied at the outside surfaces of the simulation setup, the solution of the extracellular potential in one time instance is unique except for an additive scalar constant [140, 141]. As all calculated signals are calculated as a difference between two points, this constant should be eliminated. Therefore, it was evaluated, if the Dirichlet boundary condition can be introduced in a post processing step, in which all extracellular potentials are referenced to a virtual reference potential. In order to evaluate the influence of different boundary conditions, simulations including Dirichlet boundary conditions in the corners of the simulation setup were compared to a simulation, in which this condition was omitted. In all cases, the reference electrode for unipolar EGMs was located at the upper surface of the simulation setup. Both approaches were compared regarding the signal shape of unipolar and bipolar electrograms.

Estimation of the Filter Transfer Function of the Clinical Recording System

The filter function implemented in the clinical system was estimated based on the system identification methods introduced in section 8.3. One signal with a length of 20 s was exported from the clinical system in a filtered and unfiltered version and used for estimation.

Simulation of IEGM

A detailed model of an 8F ablation catheter (*CathAblation8mm*, Fig. 9.3) was used, which was introduced in Sec. 9.1.1. Using the simulation setup described in Sec. 9.1.2, with dimensions of $42.2\,\text{mm} \times 15\,\text{mm} \times 4.4\,\text{mm}$, catheter tilt angles relative to the propagation direction (α) were varied ($0°$, $45°$, $135°$, $150°$, $165°$, $180°$). Furthermore, simulations with different tissue-electrode distances (d) were performed ($0\,\text{mm}$, $-1.2\,\text{mm}$, $5\,\text{mm}$) at a tilt angle of $90°$. Spatial resolution of the setup was $0.2\,\text{mm}$. Intracellular conductivity (σ_i) was adjusted to $0.5\,\text{S/m}$ in order to match a CV of $1.1\,\text{m/s}$. A stimulus of $90\,\text{pA/pF}$ was applied for $2\,\text{ms}$ at the front face (y-z plane, x=0) of the myocardial patch in x-direction.

Clinical Signals

Filtered clinical signals (unipolar, bipolar) were acquired as described in Sec. 8.1 for different catheter orientations under physiological conditions. All signals were acquired during pacing from the CS catheter at a basic cycle length of 600 ms. Signal orientations and positions were annotated by the physician based on haptic feedback and fluoroscopic control.

Comparison of Simulated and Measured EGM

Mean templates over 20 s created by the algorithm introduced in Sec. 8.2 were used for direct comparison to simulated data. Exact catheter orientation of clinical signals was estimated from the time lag between unipolar electrograms of both EGMs. Based on this information, a suitable simulation was chosen for direct comparison. For comparison, simulated EGMs were first filtered with the clinical filter function. Comparison was performed by visual inspection followed by calculating the correlation coefficient as measure of similarity.

Influence of Catheter Orientation, Conduction Velocity, Tissue Thickness and Clinical Filtering on Simulated Electrograms

Having validated the simulation approach with clinical data a systematic simulation study was performed to study the influence of catheter orientation, CV and tissue thickness on the unipolar and bipolar EGM. Tissue thickness between 0.8 mm and 6.4 mm was simulated for orthogonal and parallel catheter orientation. Furthermore, CVs in the range of 0.38-0.99 m/s were studied.

Influence of Different Filter Settings

Using the identified filter function, different highpass and lowpass settings were investigated, which were found in literature for clinical applications [4, 6, 58]. Regarded highpass frequencies ranged from 0 Hz to 30 Hz. Lowpass cut off frequencies were varied between 100 Hz and 500 Hz. Additionally the influence of a 50 Hz notch filter with a bandwidth of 2 Hz was analyzed.

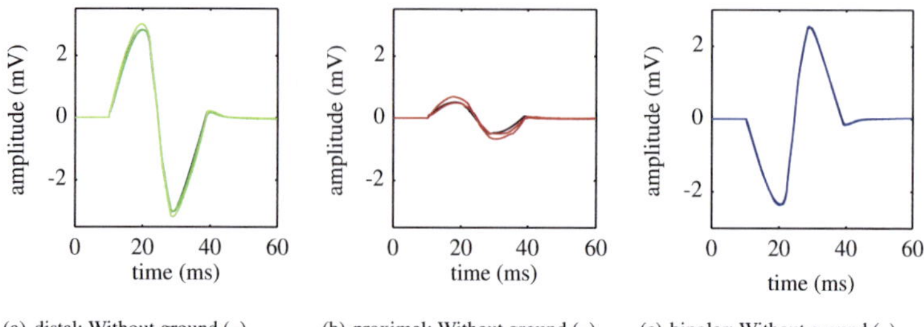

(a) distal; Without ground (–),
Upper corners grounded (–), lower
corners grounded (–)

(b) proximal; Without ground (–),
upper corners grounded (–), lower
corners grounded (–)

(c) bipolar; Without ground (–),
upper corners grounded (–), lower
corners grounded (–)

Figure 10.1. Comparison of EGMS obtained by different configurations of Dirichlet boundary
conditions

10.2 Validation of the Simulation Approach on Clinical Data

Simulations of IEGM presented in this work focus on a better understanding of morphological characteristics. This section illustrates a comparison of simulated IEGM to clinical data, which was introduced in chapter 8. Recordings from an 8F ablation catheter, with annotated catheter orientations were exported and templates over 5 s timeframes were extracted. For a direct comparison with simulated signals the filter function of the clinical acquisition system had to be estimated.

10.2.1 Evaluation of Boundary Conditions

In this section comparative results analyzing the influence of Dirichlet boundary conditions were investigated. Three simulations were compared. In one case, only no-flux Neumann boundary conditions were applied at the outside surfaces for intra- and extracellular space. Furthermore, two simulations with differently positioned ground potentials (Dirichlet conditions) were regarded. In one simulation the setup was grounded at the four upper corners. In the third simulation the ground potentials were located in the lower corners of the setup. All unipolar signals were calculated against a virtual reference surface in the blood as introduced in section 9.2. Resulting signals are shown in Fig. 10.1. Bipolar EGMs were identical for all conditions. Regarding unipolar EGMs, simulations 1 and 2 produce nearly the same result differing by less than 0.06 mV in peak-to-peak amplitude. Simulation 3 with grounded lower corners results in slightly higher amplitudes (+0.4 mV peak-to-peak) in unipolar EGMs.

10.2.2 Estimation of the Clinical Filter Function

Based on a 20 s signal, which was acquired with both, minimal filtering (highpass at 1 Hz) and the clinical standard filter (bandpass 30-250 Hz), the filter response of the clinical system was estimated as described in section 8.3. Figure 10.2(a) displays the power spectral densities of the two signals. The frequency content of both signals is limited to frequencies below 200 Hz. Therefore, a reliable estimation of the transfer function was only possible for this part of the spectrum. Among the tested standard filters, a series of a first order Butterworth highpass and lowpass have provided the closest match with the estimated transfer function. This is true, as well for the magnitude (Fig. 10.2(d)) as for the phase response (Fig. 10.2(e)). For validation, 80 unfiltered EGMs were filtered with the estimated filter and correlated with their version filtered by the clinical system. The mean correlation coefficient was 0.994 ± 0.013. Regarding an example of one set of EGMs before and after filtering, no difference between the two EGMs can be distinguished by visual inspection (Fig. 10.2(c), Fig. 10.2(b)).

10.2.3 Comparison of Clinical and Simulated IEGM

In the following, simulated EGMs were compared to clinical signals for different catheter orientations. Furthermore, the influence of changing electrode tissue distance was investigated.

EGM amplitude and morphology depend on the orientation (angle) and distance of the catheter relative to the myocardial surface, the propagation direction and velocity of the spread of excitation. In the following, EGM templates extracted from 5 s frames of clinical recordings with stable catheter position are compared to filtered simulated EGMs. The shown exemplary

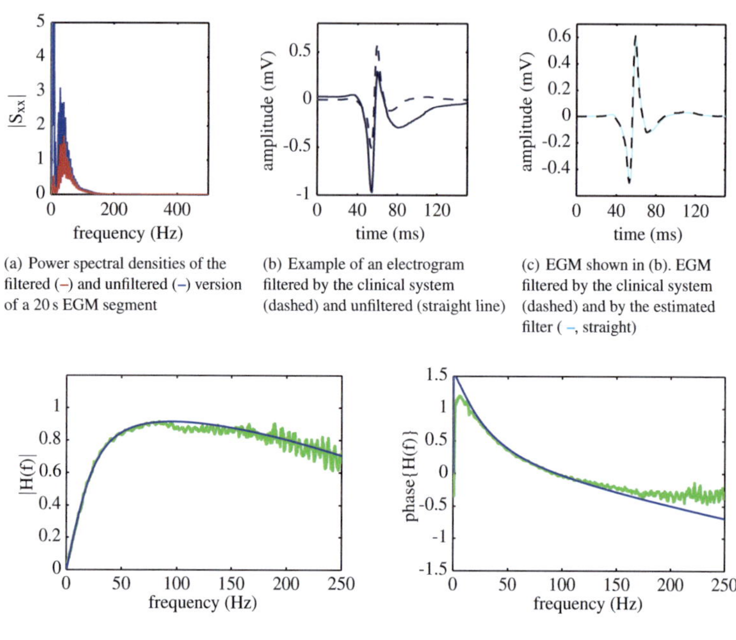

(a) Power spectral densities of the filtered (–) and unfiltered (–) version of a 20 s EGM segment

(b) Example of an electrogram filtered by the clinical system (dashed) and unfiltered (straight line)

(c) EGM shown in (b). EGM filtered by the clinical system (dashed) and by the estimated filter (–, straight)

(d) Magnitude response of the clinical system. Estimated transfer function (–). Transfer function of the estimated Butterworth filters (–)

(e) Phase response of the clinical system. Estimated transfer function (–). Phase response of the estimated Butterworth filters (–)

Figure 10.2. Estimation of the filter function used in the clinical recording system from a filtered and unfiltered signal

cases were used to provide a morphological comparison between simulation and measured data. Based on the annotation given by the physician, simulated signals were chosen, which fit the time lag of the measured unipolar EGM as well as amplitude symmetry of the bipolar EGM. For all cases the unipolar recordings of the tip of the ablation catheter and the first ring electrode are displayed with the corresponding bipolar EGM. For easier visual comparison, some of the shown EGM sets were aligned in time. Therefore, the positive peak of the bipolar EGM was used as a reference and scaled to an amplitude of 1. The unipolar data belonging to this set were shifted and scaled by the same values. This was done to keep amplitude and time relationships between the leads.

Electrode-Tissue Distance

Besides the haptic feedback, signal morphology is an indicator for the physician to judge whether the catheter is in contact with the myocardium. Presented clinical signals, in which the catheter approaches the wall, gets into contact and is finally penetrating, showed an increase in EGM amplitude and peakedness. EGM width, however, is hardly altered (Fig. 10.3(d-f)) during this process. This behavior was closely reproduced by a sequence of simulations, where the catheter initially had a distance of 5 mm to the tissue surface, secondly touchet the surface and was finally deforming the tissue surface by 1.2 mm ((Fig. 10.3(a-c)). Corresponding correlation coefficients are given in Tab. 10.1.

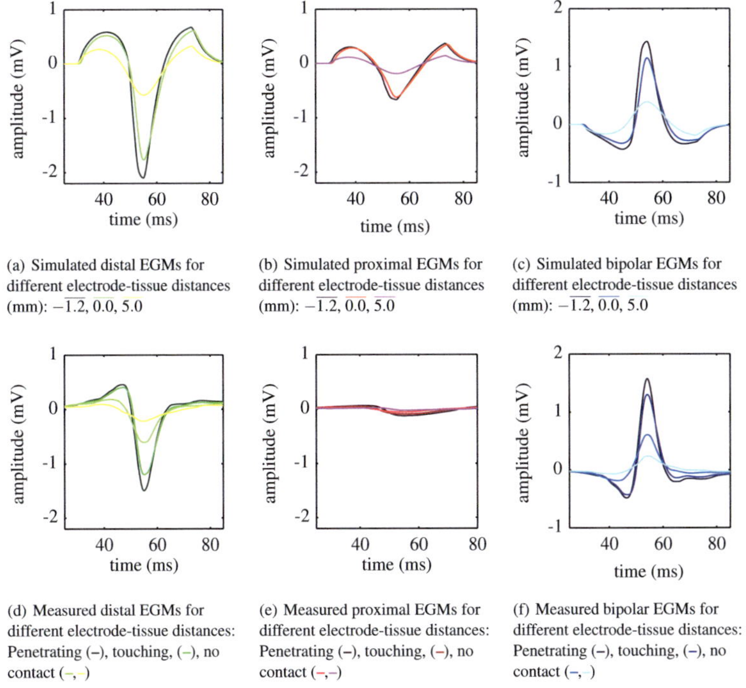

(a) Simulated distal EGMs for different electrode-tissue distances (mm): $-\overline{1.2}, \overline{0.0}, 5.0$

(b) Simulated proximal EGMs for different electrode-tissue distances (mm): $-\overline{1.2}, \overline{0.0}, \overline{5.0}$

(c) Simulated bipolar EGMs for different electrode-tissue distances (mm): $-\overline{1.2}, \overline{0.0}, 5.0$

(d) Measured distal EGMs for different electrode-tissue distances: Penetrating (–), touching, (–), no contact (–,)

(e) Measured proximal EGMs for different electrode-tissue distances: Penetrating (–), touching, (–), no contact (–,–)

(f) Measured bipolar EGMs for different electrode-tissue distances: Penetrating (–), touching, (–), no contact (–,)

Figure 10.3. Comparison of simulated and measured EGMs for different electrode tissue distances

Catheter Orientation

To investigate and compare EGM morphology with respect to catheter orientation, three sets of clinical EGM were used. The first recording position was orthogonal to the tissue surface, whereas the second ($\alpha = 135\,°$) and the third ($\alpha = 165\,°$) were oriented increasingly parallel. Shapes of measured

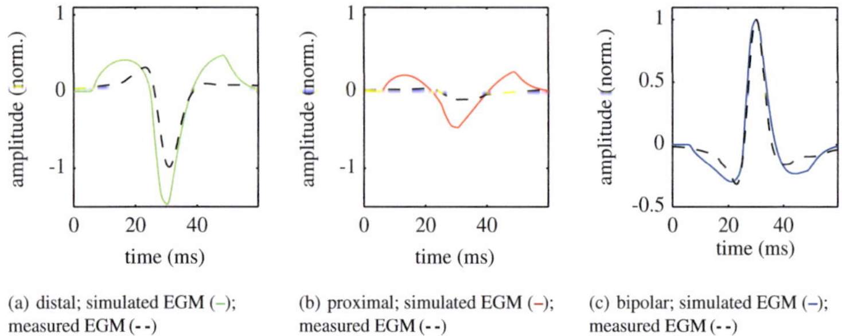

(a) distal; simulated EGM (–);
measured EGM (- -)

(b) proximal; simulated EGM (–);
measured EGM (- -)

(c) bipolar; simulated EGM (–);
measured EGM (- -)

Figure 10.4. Comparison of measured and simulated EGMs for orthogonal catheter orientation

bipolar EGMs for an orthogonal catheter orientation were in good agreement (Fig. 10.4) with simulated data. Also the morphologies of UEGM match the measured signals. However there is a difference in UEGM amplitudes. The best correspondence exists for the center part of the EGM, which coincides with the wavefront passing the electrodes. Similarities between measured and simulated EGM were quantified by correlation coefficients shown in table 10.1. In the shown examples CV was 1.1 m/s, which led to a high agreement between measured and simulated EGMs. As shown in the following parts of this chapter EGM width is dependent on this parameter and therefore needs to be adjusted in order to achieve a good fit.

Figure 10.5 displays a comparison for a parallel orientation of the catheter. Bipolar signal morphology of simulated EGM is well reproduced by the simulation. UEGM amplitude differences are smaller than for the orthogonal case and become very small for a catheter orientation of $\alpha = 165\,°$. In the proximal UEGM a small 'dip' in the simulated curves is visible. Correlation coefficients for the shown curves are given in in table 10.1. For these

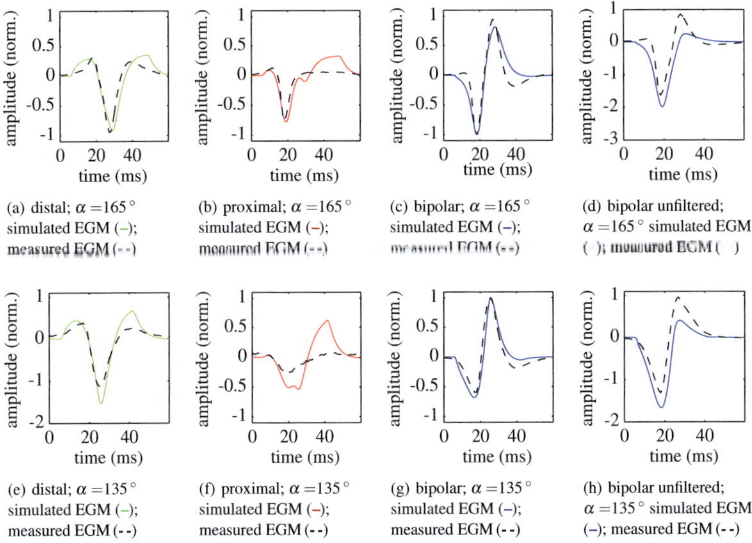

Figure 10.5. Comparison of measured and simulated EGMs for parallel catheter orientation

recordings also unfiltered bipolar EGMs were acquired, which are displayed in the fourth column of Fig. 10.5. Besides a good morphological agreement, an offset in amplitude between the measured and simulated curves is present.

Quantitative Comparison by Correlation

Correlation coefficients (Tab. 10.1) for bipolar EGM and distal UEGM are above 0.9 for all catheter orientations. For the proximal UEGM values above 0.7 with a maximum of 0.87 for a catheter angle of $165\,^\circ$ were obtained.

Table 10.1. Correlation coefficients between measured EGMs and corresponding simulation results. Values in brackets indicate the electrode-tissue distance in the simulation.

catheter position	distal	proximal	bipolar
90° (5 mm)	0.91	0.75	0.98
90° (0 mm)	0.92	0.77	0.98
90° (-1.2 mm)	0.87	0.72	0.92
135° (0 mm)	0.96	0.77	0.94
165° (0 mm)	0.93	0.87	0.92

10.3 Simulation Study of Influencing Parameters on EGM Morphology

Using the verified simulation approach, in the following different parameters, which affect EGM morphology are investigated. These parameters include different catheter orientations and electrophysiologic parameters.

10.3.1 Influence of Tip Electrode Length

Besides ablation catheters with an 8 mm tip electrode for large ablation lesions, also 4 mm catheters are frequently used. Simulation results for both geometries (*CathAblation4mm*, *CathAblation8mm*) are depicted in In Fig. 10.6. Simulated EGMs for the 4 mm catheter have a 2-3 times higher amplitude. Furthermore, the steepness of the negative deflection of unipolar EGM and the positive deflection of bipolar EGM is increased.

10.3.2 Catheter Orientation

Catheter orientation influences EGM morphology due to varying electrode positions relative to the tissue surface and a changing contact surface between electrode and tissue. Morphological changes for variation of angle α with $\beta = 0$ are shown in Figure 10.7. Besides minor changes in amplitude, distal UEGMs do not change (Fig. 10.7(a)). Proximal electrode position and distance to the tissue surface changes gradually with varying angle α. A change in amplitude symmetry can be observed. Positive and negative UEGM amplitudes are symmetric for $\alpha = 90°$. For angles greater than $90°$

the proximal electrode is located closer to the stimulation point and therefore has a more pronounced negative peak (S-wave). The reverse effect occurs for angles greater than 90°. Small double peaks are present for orientations where the distal electrode shadows the proximal electrode and leads to field distortions.

BEGM morphology changes from a monophasic negative peak (S-wave) over a biphasic shape to a monophasic positive peak (R-wave), when the catheter angle is varied between 180° and 0°. Peak to peak amplitude and symmetry strongly depends on catheter orientation. This aspect is depicted in Fig. 10.8 in which both α and β were studied, which represents the full

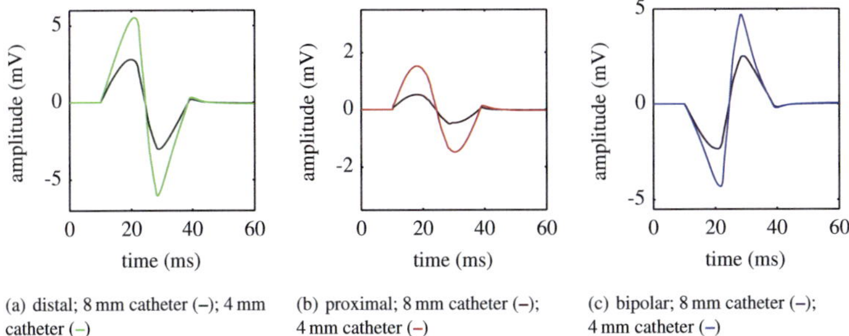

(a) distal; 8 mm catheter (–); 4 mm catheter (–)

(b) proximal; 8 mm catheter (–); 4 mm catheter (–)

(c) bipolar; 8 mm catheter (–); 4 mm catheter (–)

Figure 10.6. Comparison of simulated EGM from a 4 mm and an 8 mm tip electrode ablation catheter

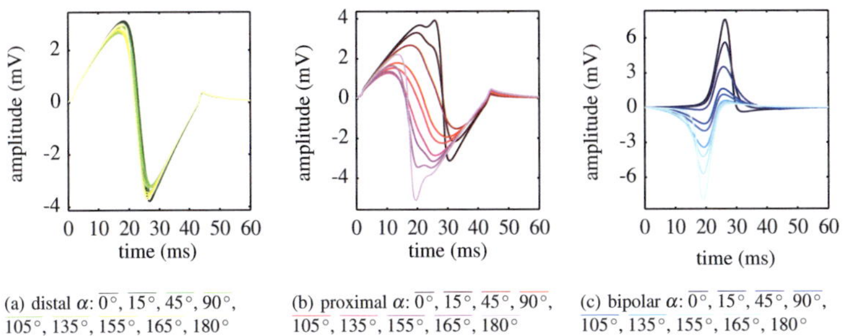

(a) distal α: $\overline{0°}$, $\overline{15°}$, $\overline{45°}$, $\overline{90°}$, $105°$, $135°$, $155°$, $165°$, $180°$

(b) proximal α: $\overline{0°}$, $\overline{15°}$, $\overline{45°}$, $\overline{90°}$, $105°$, $135°$, $155°$, $165°$, $180°$

(c) bipolar α: $\overline{0°}$, $\overline{15°}$, $\overline{45°}$, $\overline{90°}$, $105°$, $135°$, $155°$, $165°$, $180°$

Figure 10.7. Simulated EGM for different values of α

range of catheter orientations relative to the propagation direction. Maximum peak-to-peak amplitudes were found in positions where both electrodes were in contact with the myocardium and the catheter oriented perpendicular to the orientation of propagation. Smallest amplitude was seen, if the catheter was oriented parallel to the propagation front. Due to wave front curvature the two UEGM did not completely compensate each other (Fig. 10.8(a)).

Amplitude symmetry of bipolar EGMs is visualized in Fig. 10.8(b). Perfect symmetry is indicated by a symmetry value of 0, which is found for an orientation of $\alpha = 90°$. The R-peak dominates for $\alpha < 90°$. The S-peak is more pronounced in the signal for $\alpha > 90°$.

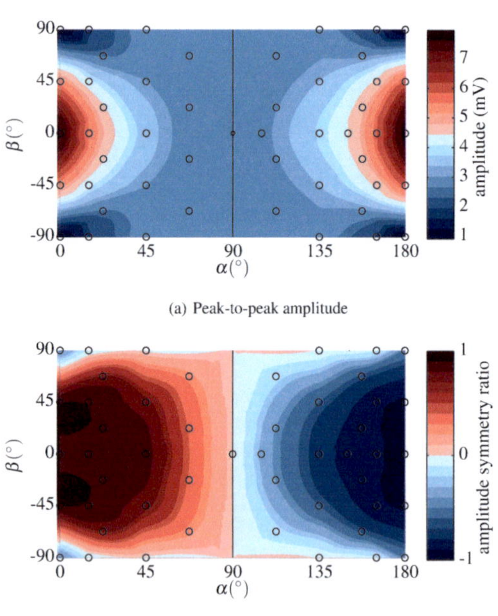

(a) Peak-to-peak amplitude

(b) Amplitude symmetry

Figure 10.8. Influence of catheter orientation on bipolar EGM morphology. α: Catheter angle relative to the myocardial surface; β: Angle relative to the propagation direction, black circles indicate simulated data points, which were used to interpolate the full plane.

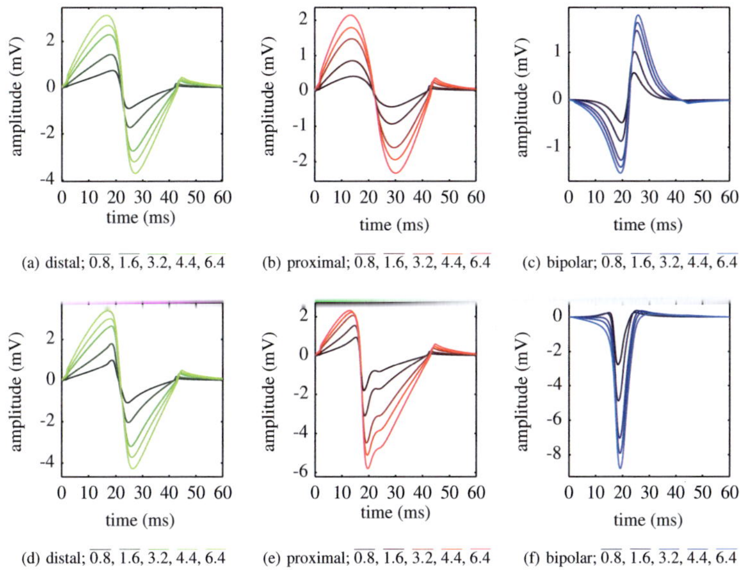

(a) distal; $\overline{0.8}$, $\overline{1.6}$, $\overline{3.2}$, 4.4, 6.4   (b) proximal; $\overline{0.8}$, $\overline{1.6}$, $\overline{3.2}$, 4.4, 6.4   (c) bipolar; $\overline{0.8}$, $\overline{1.6}$, $\overline{3.2}$, $\overline{4.4}$, $\overline{6.4}$

(d) distal; $\overline{0.8}$, $\overline{1.6}$, 3.2, 4.4, 6.4   (e) proximal; $\overline{0.8}$, $\overline{1.6}$, $\overline{3.2}$, 4.4, 6.4   (f) bipolar; $\overline{0.8}$, $\overline{1.6}$, $\overline{3.2}$, $\overline{4.4}$, $\overline{6.4}$

Figure 10.9. Simulated EGMs for different values of myocardial wall thickness (mm) with orthogonal catheter orientation, $\alpha = 0\,^\circ$ (a-c) or parallel catheter orientation, $\alpha = -90\,^\circ$(d-f)

10.3.3 Tissue Thickness

The amount of active myocardial cells contributing to the IEGM at one time instance is dependent on myocardial wall thickness. Figure 10.9 depicts EGMs simulated on patches with a thickness of 0.8-6.4 mm for orthogonal and parallel catheter orientation. Similar changes were identified for both catheter orientations. With increasing wall thickness (D), EGM amplitude increases. Furthermore, a slight increase in EGM width could be observed, while EGM shape was preserved. Quantitative analysis of the peak-to-peak amplitude showed a rapid increase for thin patches approaching an upper limit for patches thicker than 6.4 mm (Fig.10.10). Polynomial formulations and fit coefficients used to extrapolate the curves in Fig. 10.10 are given in the appendix (Tab. A.1).

10.3.4 Conduction Velocity

The influence of the speed of myocardial excitation was evaluated by varying
the CV, within a physiological range, between 0.38-0.99 m/s. The following
phenomena could be observed for orthogonal and parallel catheter orientation with decreasing CV:

- EGM amplitude decreased
- EGM width increased
- Local activation times increased

These changes occurred for unipolar and bipolar EGMs (Fig. 10.11). Besides these scaling effects EGM morphology stayed unchanged for all values of CV. The described relationships were quantified and can be seen in
Fig. 10.12. Signal peak-to-peak amplitude (V_{pp}) increased approximately by
a factor of 3 when comparing CVs of 0.5 m/s and 1.0 m/s. Parallel catheter
orientation compared to orthogonal orientation produced stronger increase
in amplitude for UEGM as well as BEGM for increasing CV. EGM width
rapidly decreased for higher values of CV, while it approached a limit for CV
values greater than 1.0 m/s. Equations and fit coefficients used to extrapolate
the curves in Fig. 10.12 are given in the appendix in Tab. A.2 and Tab. A.3.

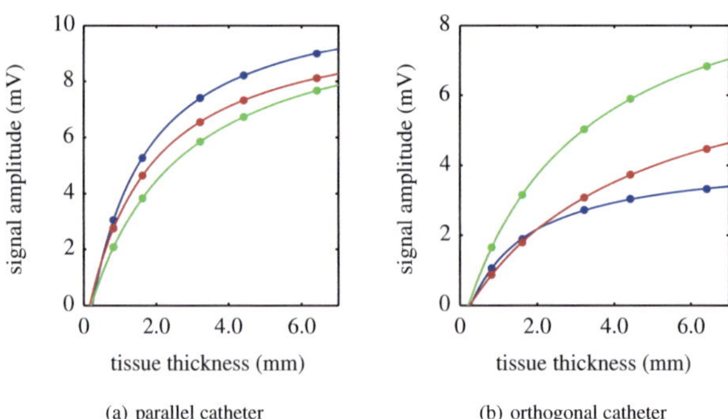

(a) parallel catheter   (b) orthogonal catheter

Figure 10.10. Peak-to-peak amplitude dependency on myocardial wall thickness. Bipolar (–),
distal (–), proximal (–)

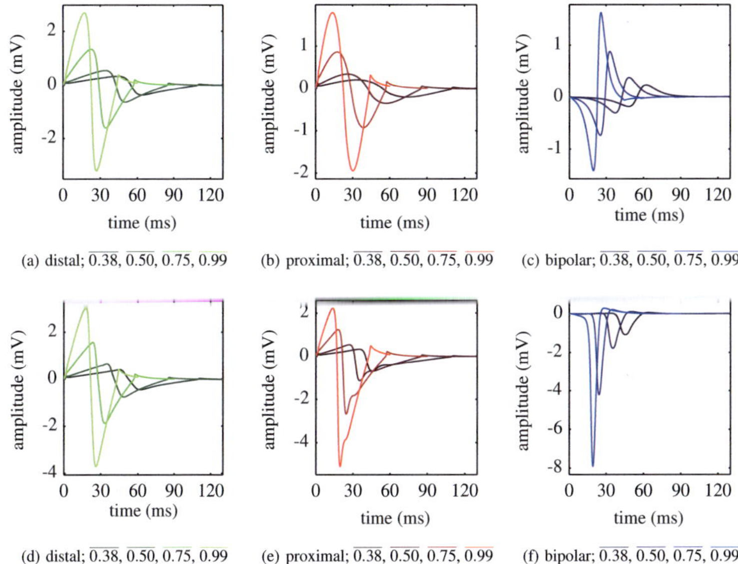

(a) distal; $\overline{0.38}$, $\overline{0.50}$, 0.75, 0.99   (b) proximal; $\overline{0.38}$, $\overline{0.50}$, $\overline{0.75}$, 0.99   (c) bipolar; $\overline{0.38}$, $\overline{0.50}$, $\overline{0.75}$, 0.99

(d) distal; $\overline{0.38}$, $\overline{0.50}$, 0.75, 0.99   (e) proximal; $\overline{0.38}$, $\overline{0.50}$, $\overline{0.75}$, 0.99   (f) bipolar; $\overline{0.38}$, $\overline{0.50}$, $\overline{0.75}$, 0.99

Figure 10.11. Simulated EGMs for different values of conduction velocity (m/s) with orthogonal (a), (b), (c) and parallel (d), (e), (f) catheter orientation

10.3.5 Influence of Different Filter Settings

As outlined previously, the clinical filter transfer function was estimated to compare simulated to measured electrograms. The identified filter function consisting of two first order Butterworth filters was used to study the impact of different cutoff frequencies on EGM morphology. For this purpose, highpass and lowpass filtering for different cutoff frequencies as well as a notch filter at 50 Hz were applied separately. Highpass cutoff frequencies ranged from 0.1-30 Hz. Lowpass cutoff frequencies were varied between 500 Hz and 100 Hz. Again, orthogonal and parallel catheter orientation were studied.

Filter Related Changes in EGMs with Orthogonal Catheter Orientation

Highpass filtering emphasized the steep parts of the EGM related to high frequency components. In both UEGM the positive deflection was attenuated and the steepness of negative deflection increased. Moreover, a second positive deflection was created at the end of the EGM. These changes led to a gradual transition of the biphasic morphology of the UEGM to a triphasic one. In biphasic EGMs the negative deflection amplitude was reduced as a result of filtering, whereas the biphasic shape was preserved (Fig. 10.13). Lowpass filtering down to 100 Hz led to minor changes in the amplitude of distal UEGM and the BEGM. Proximal UEGM were not altered (Fig. 10.14).

Filter Related Changes in EGMs with Parallel Catheter Orientation

Analogous to the orthogonal orientation, UEGM morphology was transformed from biphasic to triphasic, when highpass cutoff frequency was in-

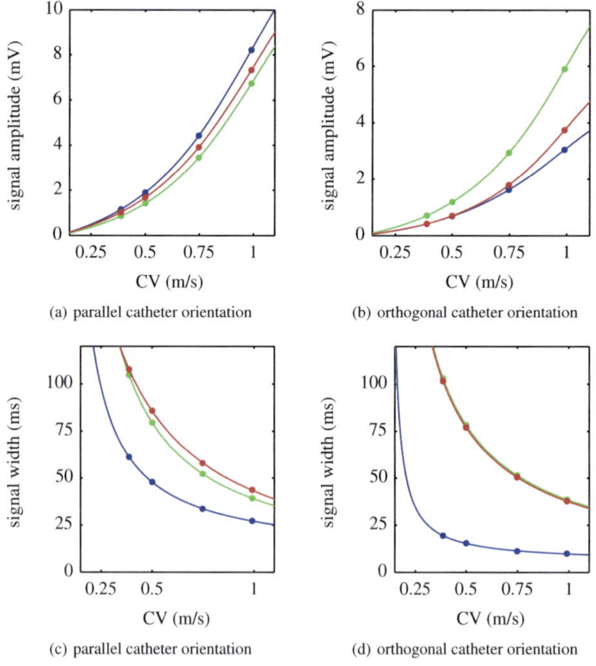

Figure 10.12. Signal amplitude (a), (b) and width dependency (c), (d) of conduction velocity. Bipolar (–), distal (–), proximal (–)

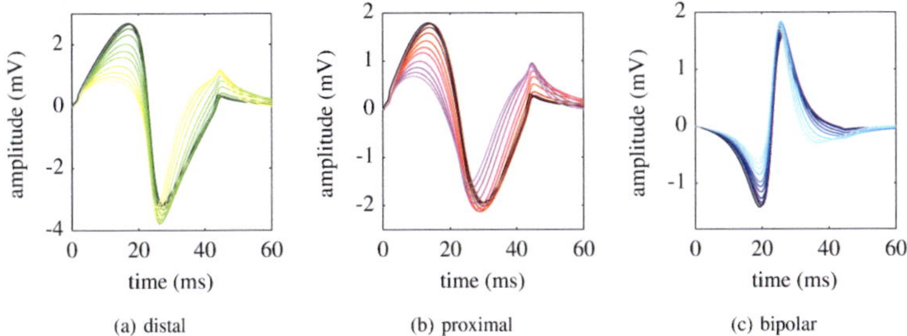

(a) distal   (b) proximal   (c) bipolar

Figure 10.13. Influence of highpass filtering (first order Butterworth) on signal morphology for orthogonal catheter orientation. Color code from dark to light colors. Cutoff frequencies (Hz): 0.1, 1, 2.5, 5, 7.5, 10, 15, 20, 25, 30

creased. BEGMs were monophasic for the unfiltered case. Highpass filtering gradually produced a biphasic signal morphology by causing a second deflection (Fig. 10.15). The applied lowpass filter did not cause significant changes in EGM morphology for parallel catheter orientation. BEGM amplitude slightly decreased. Furthermore, a small reduction in steepness of the negative deflection of the UEGM is notable (Fig. 10.16).

Notch Filtering

The bipolar EGMs in Fig. 10.17 give an impression how a quite sharp notch filter (bandwidth: 2 Hz) at 50 Hz can influence signal morphology. For the case of orthogonal orientation the negative deflection was strongly attenu-

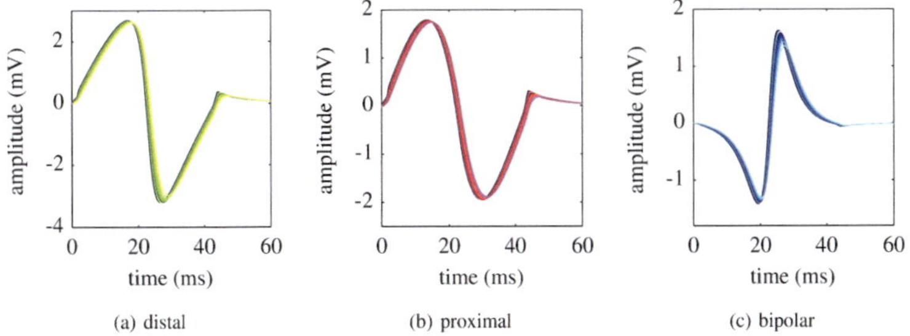

(a) distal   (b) proximal   (c) bipolar

Figure 10.14. Influence of lowpass filtering (first order Butterworth) on signal morphology for orthogonal catheter orientation. Color code from dark to light colors. Cutoff frequencies (Hz): 500, 480, 460, ... 100

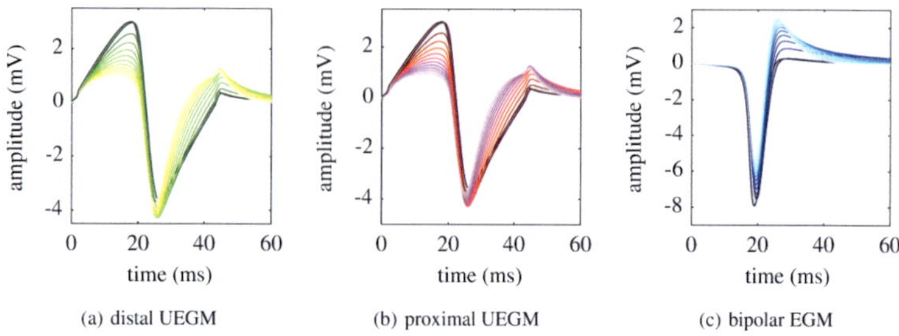

Figure 10.15. Influence of highpass filtering (first order Butterworth) on signal morphology for parallel catheter orientation. Color code from dark to light colors. Cutoff frequencies (Hz): 0.1, 1, 2.5, 5, 7.5, 10, 15, 20, 25, 30

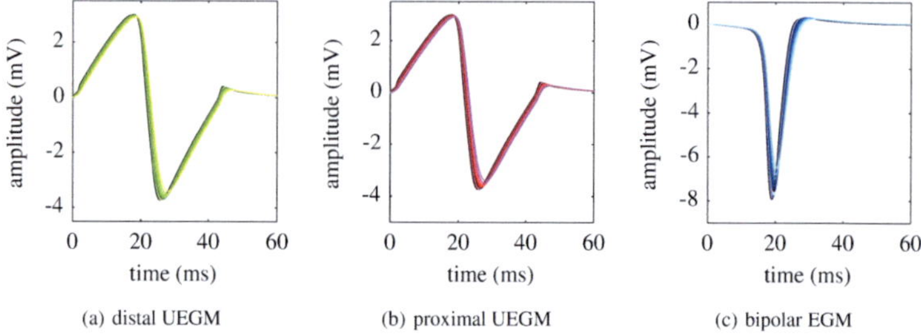

Figure 10.16. Influence of lowpass filtering (first order Butterworth) on signal morphology for parallel catheter orientation. Color code from dark to light colors. Cutoff frequencies (Hz): 500, 480, 460, ... 100

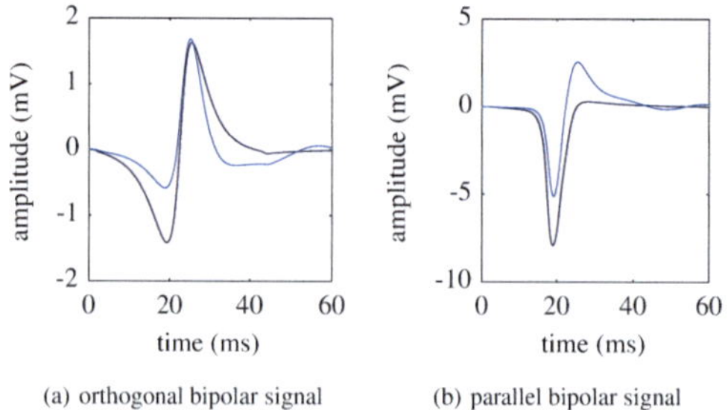

Figure 10.17. Influence of a 50 Hz notch filter with a bandwidth or 2 Hz on bipolar signal morphology. Unfiltered (–), filtered (–)

ated. If the catheter was placed parallel the signal morphology changed from monophasic to biphasic.

10.4 Discussion

Analysis of intracardiac EGMs is the state of the art in diagnosis and treatment of cardiac arrhythmias. Whereas in the past these signals were used to create annotations such as LATs, more recent studies have emphasized the diagnostic value of EGM morphology [6, 30, 142]).

Simulation studies are an ideal mean to achieve a deeper understanding of the representation of pathological substrate changes in the electrogram. Published simulation studies on clinical IEGMs can be categorized in whole atria or ventricle simulations [117, 143–145], and small scale patch simulations [28, 116, 146, 147]. The scope of whole organ simulations was the influence of global excitation patterns on IEGM morphology. These simulations analyzed forward calculated EGMs from point like sensors, neglecting the catheter geometry. Small scale simulations addressed the influence of disk electrodes [28] and furthermore, incorporated the interaction between intracellular and extracellular space by bidomain simulations [146, 147].

The study presented in this work extended the spectrum, by regarding a realistic three dimensional catheter model, which includes highly conductive electrodes and insulations between the electrodes. Using this approach a detailed verification by comparison to clinical signals, which were recorded with an identical catheter, was possible. As an additional aspect, the filter function of the clinical system was identified and used to study the impact of different filter settings. The presented simulation setup provided a thorough basis to study several influencing parameters relevant for clinical data interpretation. These parameters were catheter orientation, conduction velocity, tissue thickness and electrode-tissue distance.

Validation of Chosen Boundary Conditions

In this work, in all simulations only no-flux boundary conditions were applied at the outside surfaces of extra and intracellular space, as well as at the border between bath and tissue in the intracellular domain. This approach leads to unique solutions for extracellular potentials except for an additional scalar constant [140, 141]. Instead of applying a Dirichlet boundary condition in the simulation, this was done in a post processing step, in which all potentials were referenced to the potential of one defined reference surface. Using this approach, simulation times could be reduced and field distortions in the vicinity of ground potentials could be avoided. In the presented test simulations, bipolar EGMs were not affected by the choice of Dirichlet boundary conditions. Unipolar EGMs for the case, in which Dirichlet conditions in the upper corners were part of the virtual reference electrode, unipolar EGMs were nearly identical. When ground potentials (Dirichlet conditions) were located in the lower corners of the setup, UEGM amplitudes were shifted, which is in agreement with the findings in [140].

Verification of the Simulation Approach by Comparison of Simulated and Clinical Data

In a direct comparison to clinical EGMs a high morphological agreement for distal unipolar and bipolar EGMs was found. This was reflected by visual comparison as well as correlation coefficients above 90 % (Tab. 10.1). Also absolute amplitudes of simulated signals were in the same value range as measured EGMs (peak-to-peak amplitude around 4 mV, Fig. 10.3). Regarding the amplitudes of UEGMs and bipolar EGMs amplitude ratios of unipolar EGMs differ from measured EGMs for the cases in which electrodes are not completely in contact with the myocardium. The largest amplitude deviation for UEGM was found for the orthogonal case (Fig. 10.4), followed by a catheter orientation of 135°. Best agreement was reached for an angle of 165° (Fig. 10.5). A possible explanation may be the unknown blood conductivity, which was assumed to be 0.7 S/m in the simulation. If the hematocrit value is varied between 4% and 80% it can obtain values between 0.15 S/m and 1.5 S/m. Also the orientation of erythrocytes influences blood conductivity [129], which changes with flow velocity. Normal human hematocrit values range between 37% and 47% for women and 40-54% for men [14]. This would result in blood conductivities between 0.49 S/m and 0.75 S/m without flow.

Best agreement of measurement and simulation was achieved in the center part of the EGM, which represents the time, when the wavefront passes the catheter tip. Deviations outside this area may be related to the planar patch geometry used for the simulations. Furthermore, unipolar EGMs may be subject to differences because the reference electrode in the roof of the simulation setup is quite close compared to the WCT based reference in real measurements. Other parameters, which are not precisely known for clinical data are the conduction velocity and the propagation direction relative to the catheter.

Good morphological agreement was also obtained for unfiltered bipolar EGM (Fig. 10.5). However, a certain offset in the low pass range was present for both orientations. Several parameters may influence this lowpass component. One aspect may be the use of polarizable electrodes in the clinical case, which could not be included in the quasi static simulations.

Simulation Studies on Parameters Influencing EGM Morphology

Bipolar EGMs are the most frequently used electrode configuration in clinical practice. Due to the proximity of both recording electrodes, this configuration is less tainted with noise. In a first comparative simulation it was found, that EGMs measured by a 4 mm electrode displayed a 2-3 times higher amplitude than those of an 8 mm catheter for orthogonal catheter orientation. This effect seems to be connected to the spatial averaging effect of the electrodes. If verified in clinical or experimental data, this finding would imply, that for higher signal to noise ratios smaller electrode sizes should be used. Signal morphology was studied for different catheter orientations relative to the tissue surface (Fig. 10.7). The UEGM measured by the distal electrode, which served as a center for catheter rotation, displayed a stable morphology for all orientations. Proximal UEGMs were changing due to different electrode tissue distance and location on the simulated patch. The resulting bipolar EGM changed from a monophasic to biphasic shape for more and more orthogonal tilt. As expected the bipolar EGM turned out to be extremely dependent on orientation. Further variation of the catheter angle relative to the propagation direction (β) also led to changes in bipolar EGM amplitude and amplitude symmetry (Fig. 10.8). The clinical implication of these simulations is, that electrode orientation may be superimposed on diagnostically relevant parameters like myocardial wall thickness (Figs. 10.9), and CV (Fig. 10.12) relative to the catheter orientation. Besides amplitude changes, also EGM width was altered with changing CV. Future approaches of clinical signal analysis should include advantageous aspects of unipolar as well as bipolar EGM in order to use all available information. A promising approach is for example the determination of LAT based on unipolar EGM downstroke, which is identified by activations in the bipolar EGMs [61].

The final part of this chapter covered an analysis of different filter settings and their impact on EGM morphology. It turned out, that the 30 Hz highpass setting, which is frequently used in clinical practice strongly influences EGM morphology. Lowpass settings were investigated down to a corner frequency of 100 Hz, without significant changes in the EGM. Nevertheless, this analysis was only performed on physiological EGM. Therefore, a setting of the lowpass to higher values might still be reasonable to prevent loosing information like EGM fractionation in pathological cases. It must be stated, that clinical data is much stronger superimposed by baseline wander, therefore a higher highpass frequency can still be a good choice. However, the interpretation should always be done keeping the chosen filter setting in mind.

11

Electrical Signals Recorded by an Experimental Microcatheter

In-vitro experiments on vital cardiac tissue preparations are a unique way to study electrophysiological parameters under standardized conditions. In this chapter data from papillary muscles and atrial preparations of the rat is shown. For each preparation the excitation pattern was characterized by optical mapping (Sec. 7.2). Electrical signals were recorded by the experimental measurement sensor (Sec. 7.3.1) either as unipolar or bipolar signals. In experiments, the measurement electrode geometry, its orientation relative to the tissue surface and the position of the reference electrode are known parameters. Based on this knowledge EGMs have been simulated in an equivalent simulation setup.

After an introduction of applied methods, a comparison of simulated and measured signals is presented. This section is followed by results of three experiments, in which electrical and optical data was acquired from the same preparation. Finally, a combined simulation and measurement study on EGM changes due to varying electrode tissue distance is described.

11.1 Applied Methods

Simulation Setup

An exact geometrical representation of the experimental measurement sensor was used to simulate EGMs. The sensor was placed centered onto 5-10 mm long cylindrical strands of cardiac tissue (diameter: 1 mm). The whole setup was placed inside a bath (15 mm× 25 mm× 15mm, spatial resolution: 0.1 mm), which resembled the dimensions of the experimental tissue bath. Same as in the experiments, the reference electrode was located at the bottom of the tissue bath in a distance of 15 mm from the tissue. Different

Figure 11.1. Simulation Setup resembling the experimental measurement conditions. A model of the experimental measurement electrode is positioned centered on a cylindrical strand of myocardium with length 10 mm and diameter 1 mm. The whole setup was placed in bath (15 mm× 25 mm× 15mm), filled with Krebs-Henseleit solution.

from previous other simulations in this work, the bath conductivity was set to 1.5 S/m, which equals the measured conductivity value of the used Krebs-Henseleit solution (Measurement was performed using an LF318, Tetra-Con325, Wissenschaftliche Werkstaetten GmbH Weilheim, Germany). In order to study electrogram changes due to electrode tissue distance, a set of simulations was performed, where the catheter tip was positioned with increasing distance d_z from the tissue surface (d_z: 0μm, 200 μm,...,1400 μm, 2000 μm, 2500 μm). Simulated EGMs were extracted as unipolar EGMs between the electrodes and the described reference electrode. Furthermore, bipolar EGMs were obtained from the difference of both electrode potentials.

Experiments

Electrical and optical data was acquired from a total of 3 right atria and 2 left ventricular papillary muscles of male rats (F344). For a detailed description of the used methods of preparation, see section 7.5. All data was acquired at a tissue bath temperature of $36.7\pm0.1\,$°C. Papillary muscles were stimulated at a rate of 1 Hz. In atrial preparations it was necessary to overpace the inherent frequency of the sinus rhythm. Therefore, pacing frequencies were set between 5 Hz and 5.5 Hz. For all preparations, optical data was acquired using the voltage sensitive dye, di-4-ANEPPS at an interpolated sampling frequency of 1 kHz. Furthermore, extracellular EGMs were recorded by the presented experimental measurement sensor (Sec. 7.3.1). In papillary muscle preparations optical and electrical measurements were performed simultaneously and digitized at a sampling frequency of 5 kHz. In atrial preparations, first the optical data was recorded to characterize the propagation pattern. Subsequently, electrical data was acquired in several locations. Here the electrical sampling frequency was 100 kHz.

For the study of EGM changes due to varying electrode tissue distance, the measurement electrode was precisely positioned using the computer controlled micro-manipulator. As the point of contact could not be precisely determined, the point of maximum signal amplitude was used as zero distance. Starting at this point, the sensor was moved away from the tissue surface in steps of 200μm. Eight measurement sequences from three right atrial preparation were acquired.

Data Analysis

Optical data was analyzed as introduced in section 7.2.2. Activation time maps were produced to visualize activation patterns corresponding to the measured signals. The time point $t = 0\,ms$ for activation time maps was set to the time of stimulation. Active segments in recorded signals were identified by detection of the onset of the stimulus artifact, which also served as zero point for the time axis of the measured EGM plots. In order to remove high frequency noise, data was lowpass filtered using a digital fourth order Butterworth filter with a cut off frequency of 1500 Hz.

11.2 Comparison of Measured and Simulated Signals

Simulations and measurements in this section were performed using identical measurement electrode geometries and reference electrode positions. The electrogram was measured in the pectinated portions of atrial preparation 2. Based on the peak-to-peak amplitude development (see Sec. 11.4) the EGM of the position was chosen at which the catheter was in contact, but not deforming the tissue. In the simulation, the measurement sensor was placed centered on a cylindrical strand of cardiac tissue, with a length of 5 mm, which was approximately the distance between the stimulus location and the border of the preparation. The set intracellular conductivity of 0.5 S/m in this case resulted in a CV of 1.25 m/s. Both EGMs are shown in Fig. 11.2. Electrogram width and the ratio between positive and negative peak amplitude show high agreement. The downstroke part and the shape of the negative peak are nearly identical. Deviations can be identified in the EGM onset and the positive peak shape. The first millisecond of the signal is dominated by the stimulation artifact, which produces a pronounced upstroke. The ends of the EGMs also deviate. The simulated EGM shows a small overshoot. The measured EGM end is superimposed by a second remote activation.

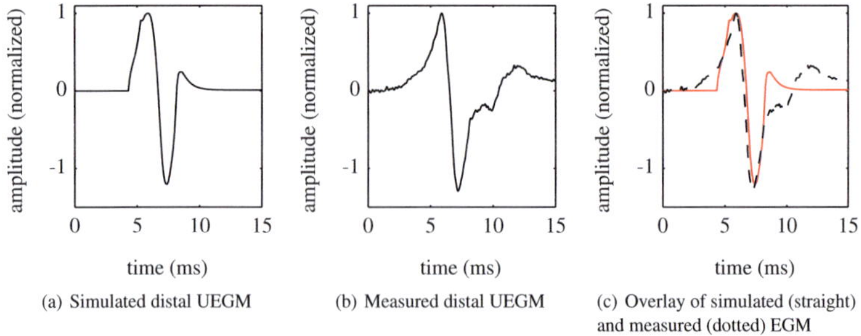

(a) Simulated distal UEGM   (b) Measured distal UEGM   (c) Overlay of simulated (straight) and measured (dotted) EGM

Figure 11.2. Exemplary measured and simulated distal UEGMs

11.3 Simultaneous Optical and Electrical Measurements

This section presents three experiments with combined optical and electrical data as case studies. First a bipolar measurement on a left ventricular papillary muscle is presented. In the following, signals and optical data of two atrial preparations is shown.

Papillary Muscle of the Left Ventricle

A transmitted light image of the examined papillary muscle is shown in Fig. 11.3(a). The image shows the endocardial side of the preparation, from which optical data was recorded. The preparation was stimulated with the tungsten needle, visible on the right side of the transmitted light image. Optical signals containing action potentials were obtained from a part of the surface area in the center of the preparation (Fig. 11.3(b)). The region with sufficient data quality was used to extract the shown activation times. From the first detected activation it took 6 ms for the wavefront to transverse the region of interest.

For electrical signal recordings, the measurement sensor was positioned on the epicardial surface of the preparation. The shown bipolar signal (Fig.11.4(b)) was recorded at the position, where the sensor was centered on the papillary muscle (Fig. 11.4(c)). Due to the close position to the stimulation and a necessary stimulus amplitude of 5 mA, the recording amplifier was saturated after the stimulus, which lead to a decreasing DC offset superimposing the bipolar signal. In order to remove this capacitive discharge curve, is was fitted by an exponential curve, which was subsequently subtracted to restore the original baseline (Fig. 11.4(b)). The resulting bipolar signal shows a biphasic negative to positive morphology, which was anticipated for the orthogonal orientation of the measurement sensor relative to the tissue surface. The time between EGM onset and electrogram end is 6 ms. This is in agreement with the optically acquired data.

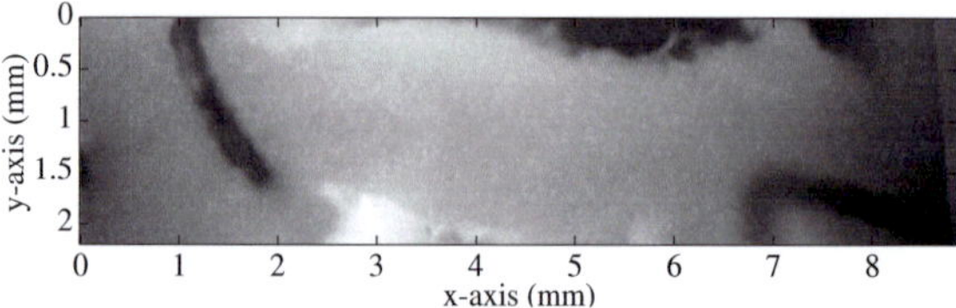

(a) Transmitted light image of a left ventricular papillary muscle. The muscle is fixed onto a silicone disc by tungsten needles (black).

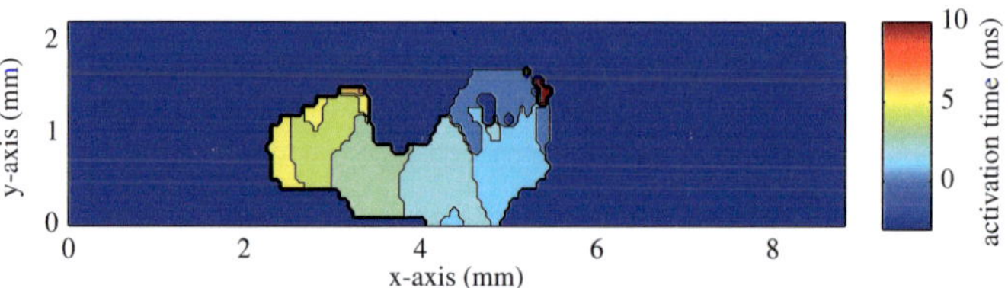

(b) Local activation times extracted from the fluorescence signal

Figure 11.3. Optical characterization of the excitation pattern on a left ventricular papillary muscle.

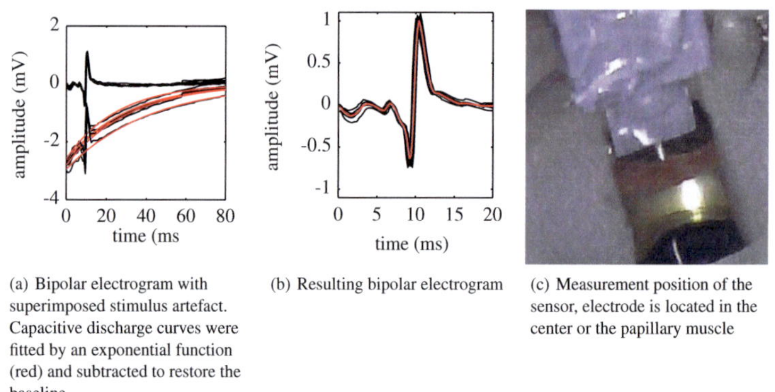

(a) Bipolar electrogram with superimposed stimulus artefact. Capacitive discharge curves were fitted by an exponential function (red) and subtracted to restore the baseline.

(b) Resulting bipolar electrogram

(c) Measurement position of the sensor, electrode is located in the center or the papillary muscle

Figure 11.4. Bipolar signal recorded on a left ventricular papillary muscle

Right Atrial Preparation 1

A photographic image of the first right atrial preparation is shown in Fig.11.5(a). The stimulus location is indicated by an asterisk. A gray rectangle indicates the field of view, which was recorded by the optical system. Distances between isochrones in the activation time map (Fig.11.5(b)) visualize the conduction velocity in different areas of the preparation. Along thicker muscle bundles larger distances indicate faster propagation. Conduction slowed down in the network of pectinate muscles in the center of the preparation. Overall, the wavefront started in the area of the stimulus location and spread radially. Further away from the stimulus, the curvature of the wavefront was reduced and ended in a nearly planar wavefront.

Recorded unipolar EGMs of this preparation are displayed in Fig. 11.6. All EGMs were recorded in the pectinated portions of the preparation. Additionally, the LATs give a further orientation on the recording location. UEGM 1.1, with an LAT of 7.8 ms was located close to the upper border of the field of view. According to the optical data (Fig.11.5(b)) this area is first activated 5 ms after the stimulus. Based on the LATs of 9.5 ms and 11.0 ms, UEGM 1.2 and UEGM 1.3 were located in the center and towards the lower border of the field of view. UEGM morphologies differed between the recording locations. UEGM 1.1 had a pronounced s-wave and a small r-wave. UEGM presented a symmetric amplitude ratio, whereas in the recording location of UEGM 1.3 a larger r-wave was obtained. When stimulation was switched off, keeping the recording electrode at the position of UEGM 1.3, the morphology of the recorded UEGM 1.4 presented a less regular shape with increased number of deflections.

Right Atrial Preparation 2

In this section a second experiment on an atrial preparation is presented. This preparation showed a unique propagation pattern due to a potential scar area in the pectinated portions of the atrium. The potential scar area is visible in the photographic image in Fig. 11.7(a). Two further observations support the assumption of a possible scar. On the one hand, conduction block was present in the optical data (Fig. 11.7(b)). On the other hand, fluorescent dye intensity was drastically higher than in surrounding myocardium. The same effect is usually observed in areas of connective tissue. The preparation was

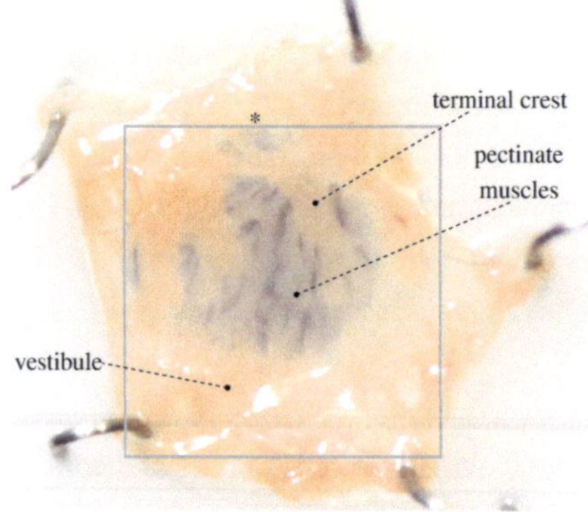

(a) Photographic image of the endocardial view of the atrial preparation. An asterisk (*) indicates the position of the stimulus electrode. The field of view of the optical system is delinated by a gray rectangle.

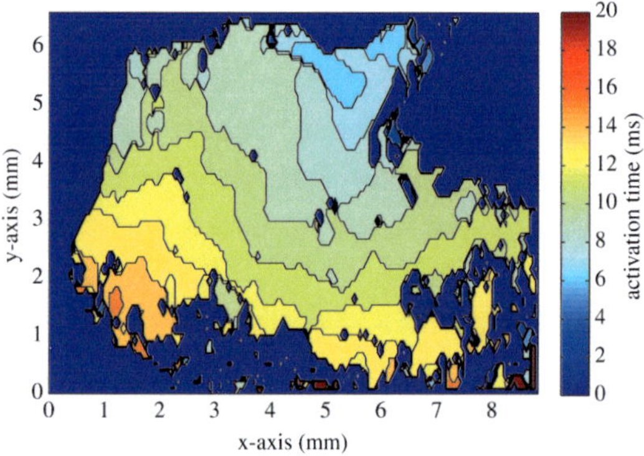

(b) Activation time map of the right atrial preparation shown in (a)

Figure 11.5. Right atrial preparation 1

stimulated near the onset of the terminal crest. From this point activation moved relatively fast along this thicker myocardial bundle. Slower conduction was present along the pectinate muscles and when traveling transversely behind the block line. The activation sequence is visualized by a time series of images, which display the fluorescent dye intensity (Fig 11.8).

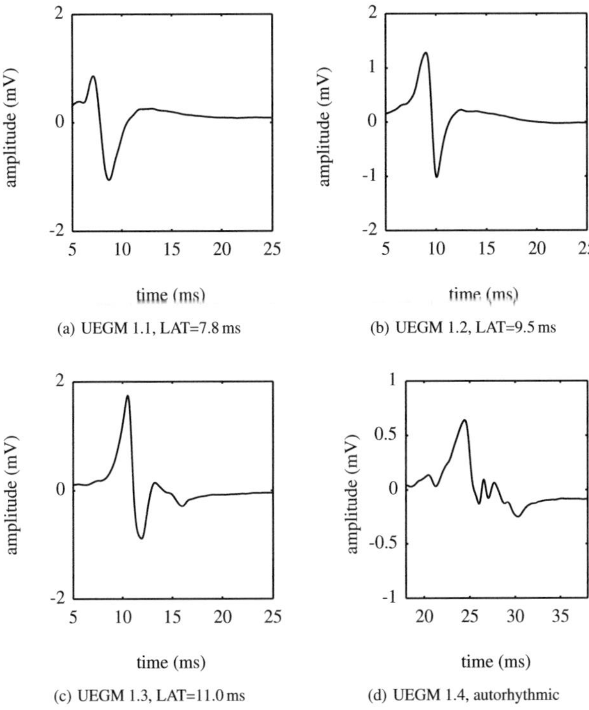

Figure 11.6. Unipolar electrograms recorded by the experimental sensor on the right atrial preparation shown in Fig. 11.5; Time axis are zeroed to the onset of the stimulation pulse. UEGMs 1.1-1.3 were recorded during stimulation, UEGM 1.4 was acquired, when during autorhytmus at the same position as UEGM 3.

Unipolar EGMs measured on this preparation are displayed in Fig. 11.9. UEGM 2.1 shows a pronounced r-wave along with an activation time of 10 ms. This data indicates a location near the border of the optical field of view (Fig. 11.7(b)). Recording locations of UEGM 2.3 and 2.4 were in the block area. Both electrograms show more than one deflection. UEGM 2.3 presents a notch in the s-wave. UEGM 2.4 displays a rather complex morphology, in which four distinct activations can be observed. Three of them with LATs at 11.8 ms, 14.1 ms and 17.7 ms are small and may arise from remote activation. The third activation coincides with an activation

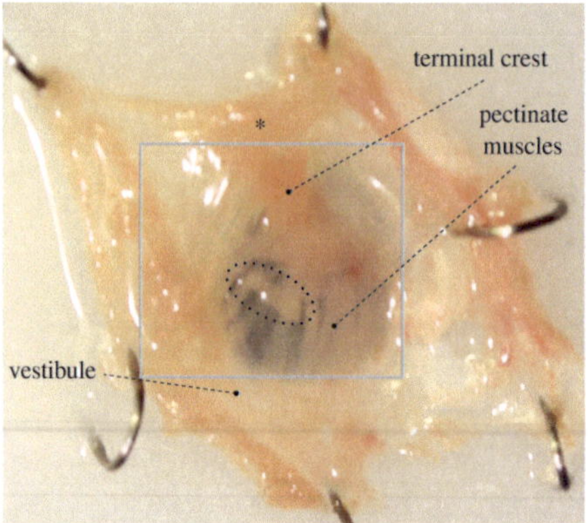

(a) Photographic image of the endocardial view of the atrial preparation. An asterisk (*) indicates the position of the stimulus electrode. The field of view of the optical system is delinated by a gray rectangle. A possible scar area is indicated by a dotted circle.

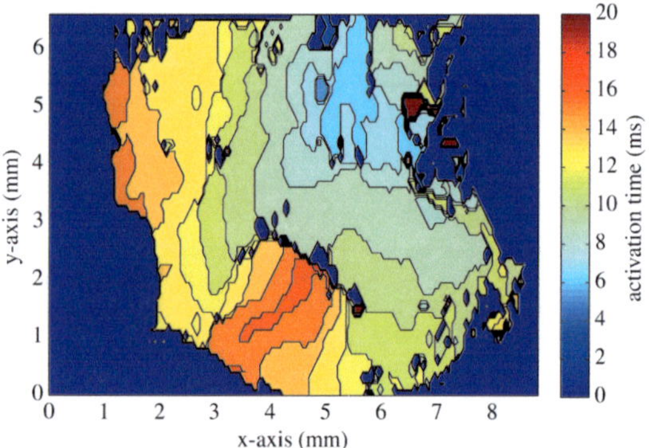

(b) Activation time map obtained from optical mapping data. Activation enters the field of view at the upper border and is blocked by the potential scar area shown in (a)

Figure 11.7. Right atrial preparation 2

time of 19.5 ms. One bipolar recording which was acquired without altering the sensor position compared to the unipolar measurement is shown in Fig. 11.10. The shown bipolar EGM corresponds to UEGM 2.2. Both EGMs display a rather monophasic shape. The bipolar EGM can be obtained from

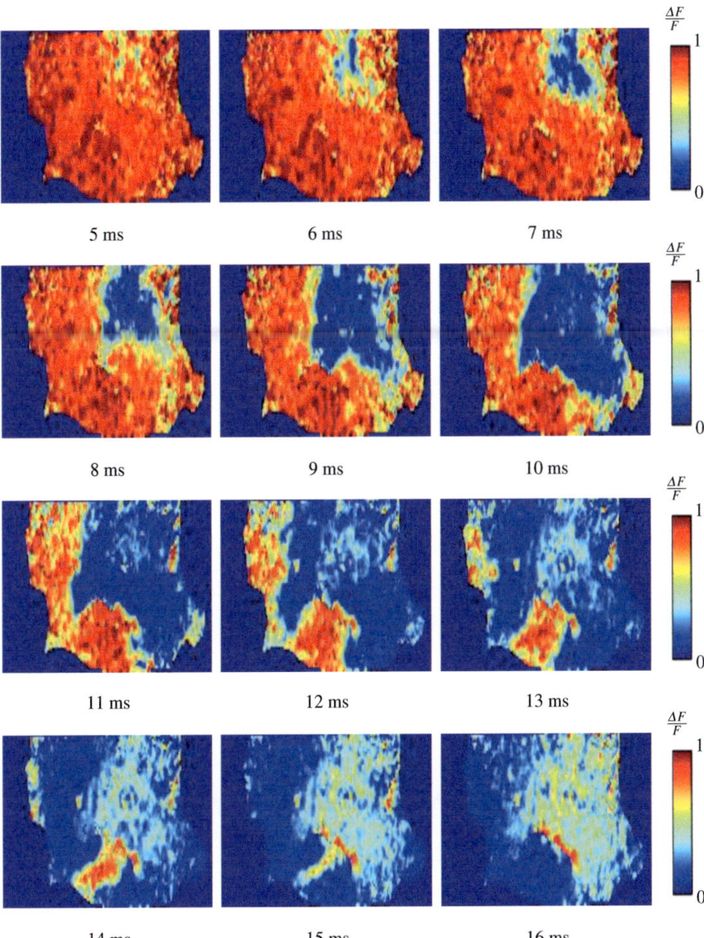

Figure 11.8. Activation sequence of the right atrial preparation displayed in Fig. 11.7. Each image shows the normalized fluorescence image of one time instant. Red indicates cells in rest, depolarized areas are colored in blue.

UEGM 2.2 by inverting the polarity. In the same recording location UEGMs without stimulation were acquired. As in the previous reported case, EGM shape changed, which indicates, that the conduction pathway under sinus rhythm conditions differed from the stimulated one.

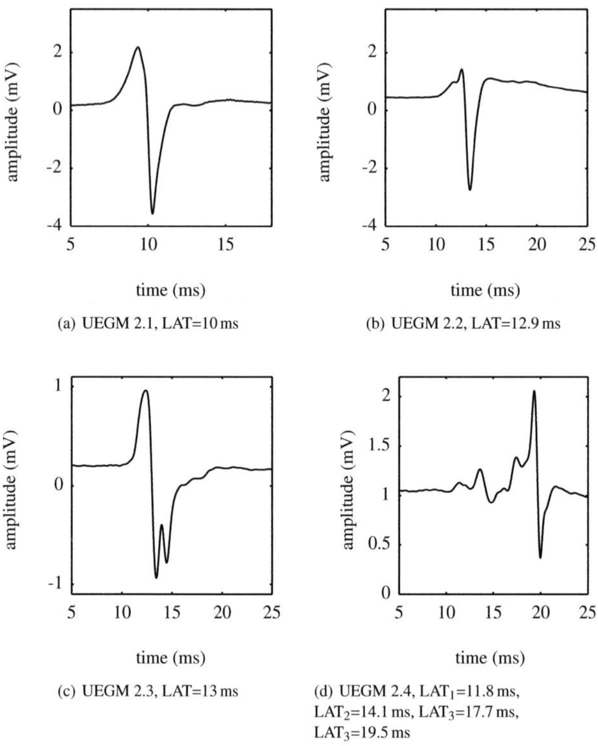

(a) UEGM 2.1, LAT=10 ms

(b) UEGM 2.2, LAT=12.9 ms

(c) UEGM 2.3, LAT=13 ms

(d) UEGM 2.4, LAT$_1$=11.8 ms, LAT$_2$=14.1 ms, LAT$_3$=17.7 ms, LAT$_3$=19.5 ms

Figure 11.9. Unipolar EGMs recorded on right atrial preparation 2

11.4 EGM Amplitude Dependency on Electrode-Tissue Distance

In the interpretation of clinical data, the detection of wall contact of measurement electrodes is an important task. To contribute to this issue, unipolar and bipolar electrograms were measured by the experimental sensor and at the same time simulated using an equivalent geometry (see Fig. 11.1). In this section the development of the peak-to-peak amplitude is investigated in relation to electrode tissue distance.

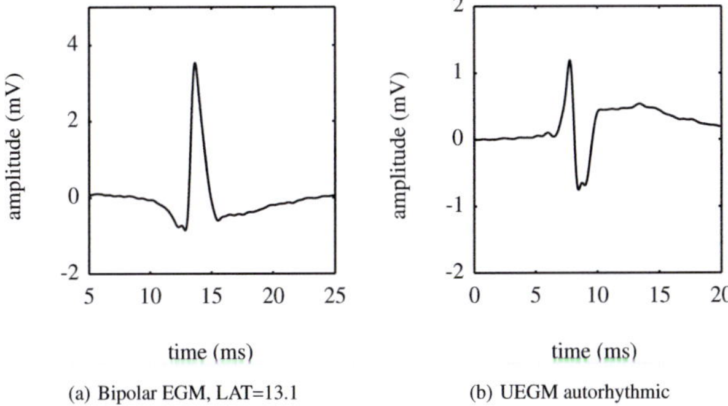

(a) Bipolar EGM, LAT=13.1 (b) UEGM autorhythmic

Figure 11.10. EGMs recorded in the same location as UEGM 2.2.; one EGM (a) was acquired in a bipolar setting between the two electrodes of the measurement sensor. The other (b) was recorded under auto-rhythmic conditions of the atrial preparation.

11.4.1 Simulated Data

A sequence of simulations with stepwise increased electrode tissue distance was performed (Fig 11.11) on a cylindrical strand of myocardium (10 mm). Peak-to-peak amplitude of distal UEGM decreased from 1.29 mV to 0.15 mV within a distance of 2.5 mm from the tissue surface. Within the same distance interval, bipolar EGM amplitudes decreased from 1.19 mV to 0.11 mV. Relative amplitude changes and peak-to-peak amplitudes relative

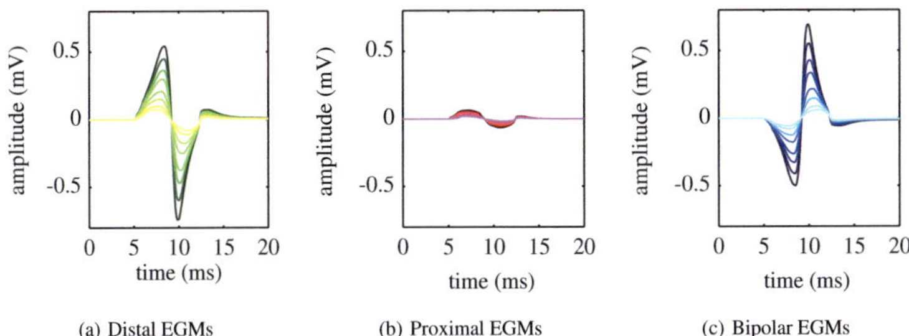

(a) Distal EGMs (b) Proximal EGMs (c) Bipolar EGMs

Figure 11.11. Simulated EGMs using the experimental measurement electrode. Distance between tissue and distal electrode was stepwise increased. From dark to light colors: 0, 0.2, 0.4, 0.6, 1.0, 1.4, 2.0 and 2.5 mm electrode tissue distance.

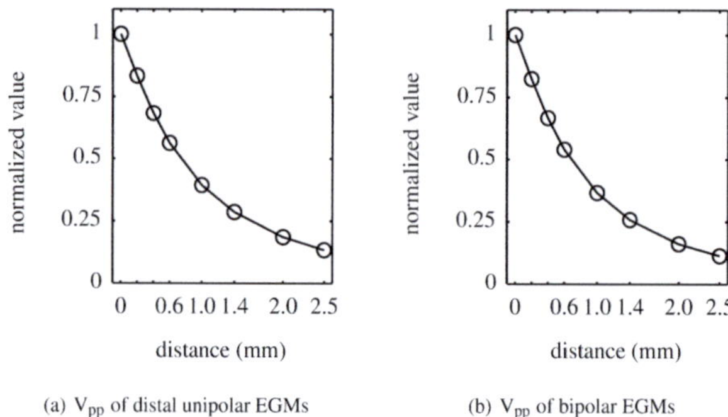

(a) V_{pp} of distal unipolar EGMs

(b) V_{pp} of bipolar EGMs

Figure 11.12. Quantitative values of the peak-to-peak amplitude (V_{pp}) relative to the electrode tissue distance in simulated EGMs for the experimental sensor.

to electrode tissue distance are displayed in Fig. 11.12. All parameters were subject to a nonlinear decrease. V_{pp} of distal UEGM decreased by 86.5% within 2.5 mm compared to 88.5% in bipolar EGM.

11.4.2 Measured Electrograms

Exemplary measured unipolar and bipolar EGMs for increasing electrode tissue distance are shown in Fig 11.13. Similar to the simulated data, amplitudes of measured signals decreased within several millimeters. Evaluating

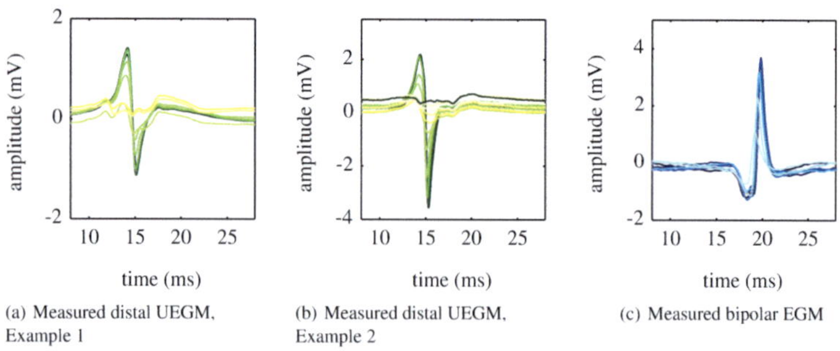

(a) Measured distal UEGM, Example 1

(b) Measured distal UEGM, Example 2

(c) Measured bipolar EGM

Figure 11.13. Experimentally acquired EGMs for increasing electrode tissue distance. Starting at the point of maximum signal amplitude, the sensor was moved away from the surface in steps of $200\mu m$. Distance increased from dark to light colors.

the peak-to-peak amplitude for each step produced a sigmoidal shaped curve, which is displayed for the signal in Fig. 11.13(a) (Fig. 11.14(a)). Within a distance of 2000 μm the amplitude decreased from 5.7 mV to 0.4 mV, which is a decrease of 93%. The amplitude did not decrease further, when the sensor was moved to a distance of 2500 μm. The displayed curve differs from the simulated data, which does not contain an inflection point. Based on the knowledge, that in simulations, the measurement electrode was not deforming the tissue surface, it was assumed, that the inflection point correlates with the point of contact with the tissue surface. Using this assumption, all eight experimental curves were aligned at the inflection point. Furthermore amplitudes were normalized to the amplitude at this point. All resulting data points are depicted in Fig. 11.14(b). The median values for each distance step are plotted in red. It was found, that when distance was increased starting at the inflection point, all measured curves show a similar decrease of peak-to-peak amplitude within a distance of 1500 μm to a threshold value near zero. Moving the sensor further in the direction of the tissue surface led to an increase in amplitude, reaching the maximum amplitude within 500 μm.

11.5 Discussion

In this chapter, extracellular electrograms, which were measured using a newly developed electrode geometry, were analyzed. Excitation patterns on the studied tissue preparation were characterized by fluorescence optical mapping data. The used experimental measurement setup combines the opportunities to acquire electrical and optical of the same preparation. Electrical signals were obtained using measurement electrodes with a similar length to catheters electrodes in clinical use. The optical recording system enables the characterization of excitation patterns. Both methods are commonly used in experimental research. However, the specific setting in the presented setup is unique. Some studies are close to the presented study in some aspects. One of the pioneering works on optical mapping of rat papillary muscles was presented in [148], where a 16x16 array of photodiodes

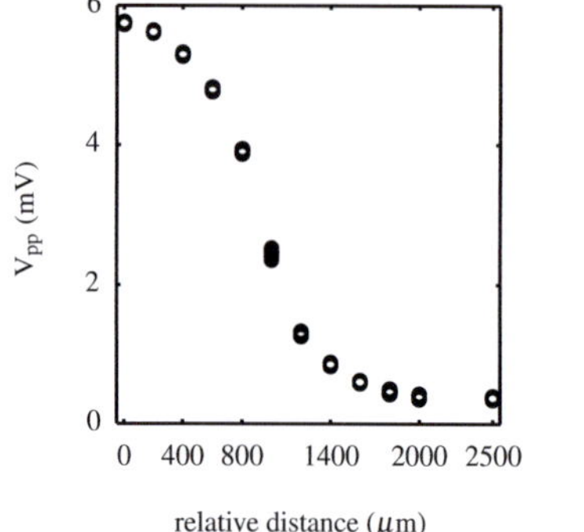

(a) Exemplary curve; each distance step contains amplitude values of 10 subsequently acquired beats.

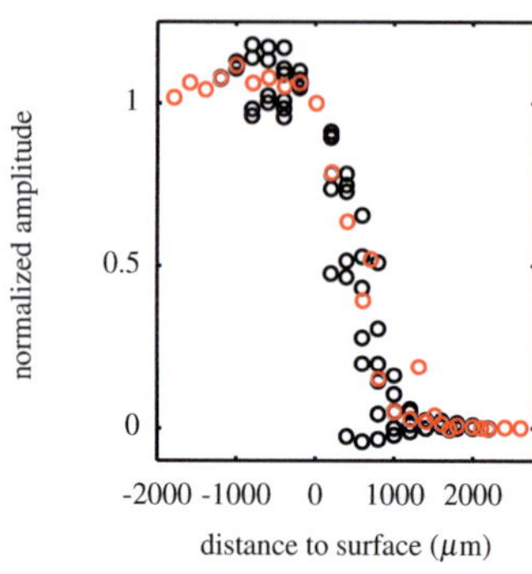

(b) Aligned and normalized data values received from 8 datasets (black). Red circles indicate the median values for each distance step. Datasets were aligned at the point of inflection and scaled to the amplitude value at that point.

Figure 11.14. Measured data on peak-to-peak amplitude development with increasing electrode tissue distance.

was used to record optical signals from the surface of this preparation. Endocardial optical mapping of superfused atrial preparations of rat and rabbit were presented in [149, 150]. Other studies investigating the rat heart by optical mapping were presented in [74, 151]. Some studies already investigated simultaneous optical and electrical recordings. A special translucent electrode grid of indium tin oxide electrodes was used, which allowed for simultaneous optical recordings [152–156]. However these extracellular sensors differ from clinical electrodes regarding size, material and spacing of the electrodes. Important studies of extracellular recordings can be found in [112, 157–159].

In the presented data, excitation patterns were characterized analyzing optical mapping data. Based on activation time maps, areas of fast and slow conduction in atrial preparations were identified. Areas of high CV corresponded to thicker muscular bundles in the area of the crista terminalis. CV in pectinated regions was reduced. The known excitation patterns supported the interpretation of measured electrogram morphologies. Comparison to simulated EGMs, which were simulated with a simulation setup of the same dimensions as the tissue bath showed good agreement. However, not all features of measured EMGs were reproduced by the simulation. This comparison shows the applicability and potential of the chosen approach for combined experimental and simulation studies. Nevertheless, deviations between simulation and experiment still express the need to reproduce the anatomy of the experimental preparation in a more accurate way.

Analyzing electrogram shapes, a correlation between the amplitude ratio of positive and negative peaks regarding the position on the preparation was observed. This finding is in agreement with previous studies [157, 160]. In the second presented atrial preparation a complex excitation pattern in the area of a scar was reflected by multiple remote activations in one unipolar EGM, which was recorded in this area. It must be emphasized, that endocardially measured excitation patterns in atrial preparations may differ from the excitation on the epicardial side, where the electrical signals were measured. However, due to the epicardial stimulation position a similar pattern can be assumed. Future studies have to reveal, to which extend the two sides agree.

In the last section of this chapter, a combined modeling and experimental study on the influence of electrode tissue distance was presented. The main finding of the presented study was a sigmoidal shaped curve of peak-to-peak amplitude over distance. It was found, that the major decrease in amplitude occurred within a distance of 1500 μm from the inflection point of the curve, which was assumed to coincide with the contact point of the electrode with the tissue surface. Moving further towards the tissue surface led to an increase in amplitude, reaching the maximum at a distance of 500 μm from the inflection point.

A similar study was already performed in 1973 [158] by Spach et al., where extracellular potentials were measured by glass microelectrodes [158]. Herein, the major amplitude decrease was reported to occur within a distance 500μm from the tissue surface, which is smaller than the simulated and measured values of the study in this work. However, the shape of decreasing curve was similar to the results found in the presented study. This data also supports the assumption, that the inflection point of the sigmoidal curve coincides with zero electrode tissue distance, as the curves shown by Spach et al. also lack an inflection point. In comparison to this, the presented study in this work produced similar results regarding the peak-to-peak amplitude decrease when moving away from the tissue surface. Furthermore, it added the part of the curve, in which an extracellular electrode is indented into the tissue, which is the common recording position in clinical practice.

12

Influence of a Measuring Electrode on Cellular Characteristics

The presence of a catheter electrode with a size of several millimeters leads to a large change in extracellular conductivity next the myocardial surface. It has been reported, that the high conductivity of blood affects the transmural shape of the excitation front [161]. Based on this fact an even higher impact of the catheter electrode can be assumed. Conceivable effects may range from higher CV and averaging effects on extracellular potentials to changed transmembrane voltage. Moreover, cells adjacent to the catheter may even be unexcitable due to the high electrical load imposed by the measurement electrode.

These aspects were addressed comparing simulations including highly conductive electrodes to reference simulations without electrode. Thereby, the influence of a catheter electrode on conduction velocity, extracellular potentials and transmembrane voltages was studied.

12.1 Applied Methods

Two types of electrodes were used to study electrode related effects on electrophysiological parameters. An ablation catheter electrode as shown in Fig. 9.3 with a diameter of 2.3 mm was placed on a myocardial patch (50 mm×6 mm×3 mm) to study EGM characteristics and changes in transmembrane voltages and extracellular potentials with a spatial resolution of 0.2 mm. In order to achieve a higher spatial resolution for the study of

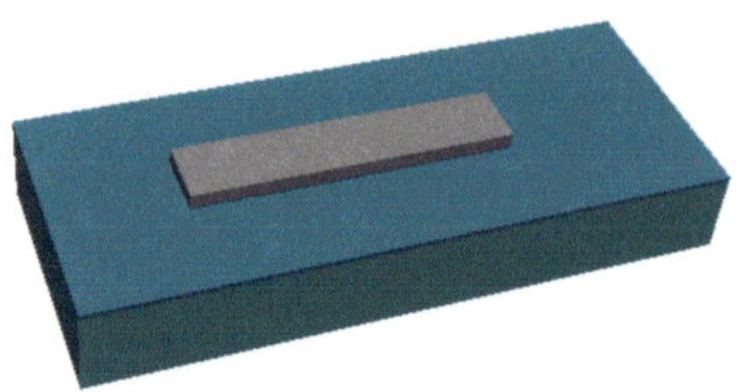

Figure 12.1. Test-setup to study electrode related changes in simulated electrophysiological parameters. A plate electrode (20 mm×6 mm×1 mm) is placed on a patch of myocardium (40 mm×15 mm×3 mm) in order to enhance electrode effects. The entire setup is surrounded by blood

CV in the electrode area, a voxel size of 0.06 mm was used on a shorter patch (12 mm×9 mm×3 mm). Additionally, a test setup with a large plate electrode (20 mm×6 mm×1 mm) was used to emphasize electrode related changes (Fig. 12.1). In the latter case, a myocardial patch of 40 mm×15 mm×3 mm was included. Comparison of transmembrane voltage and extracellular potentials per time step was performed by subtracting the reference simulation from the simulation including the catheter. Furthermore, LATs were derived from the upstroke of the action potential and again compared by subtraction. Finally, the distribution of extracellular currents was studied in the presence of the large plate electrode. Extracellular currents were calculated as introduced in section 9.2.3

12.2 Influence of a Catheter Electrode on Electrophysiological Characteristics

A catheter electrode adjacent to the myocardial surface acts as an area of high conductivity, which changes the load situation imposed onto the cardiac cells. In the following the influence on transmembrane voltages, extracellular potentials, extracellular currents and conduction velocity is described.

12.2.1 Transmembrane Voltages and Extracellular Potentials

The time series displayed in Fig. 12.2 shows, that the influence of the catheter electrode is not limited to the electrode area, but influences the electric field in a distance of several millimeters. When the wavefront entered the electrode area, the spatial averaging character limited the rise of the extracellular potential, which caused a negative difference at the right edge of the catheter. Similar effects can be seen when the wavefront passes the electrode. Field differences were most pronounced around the catheter tip. Corresponding catheter signals are displayed in Fig. 12.3. Signals were obtained from extracellular potentials in the location of the catheter electrodes. EGM amplitudes were smaller by a factor of two in the simulation including the catheter.

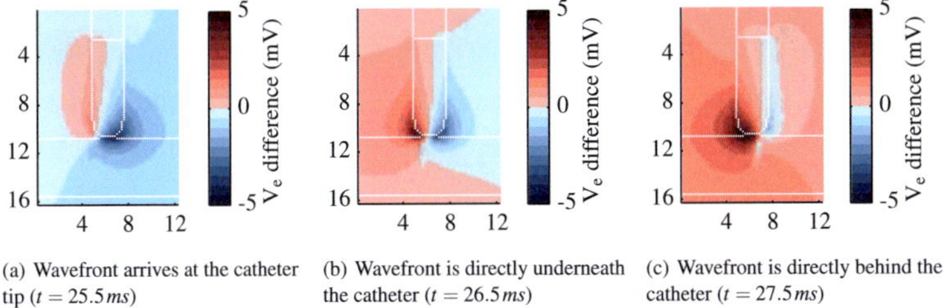

(a) Wavefront arrives at the catheter tip ($t = 25.5\,ms$)

(b) Wavefront is directly underneath the catheter ($t = 26.5\,ms$)

(c) Wavefront is directly behind the catheter ($t = 27.5\,ms$)

Figure 12.2. Time series of the difference in extracellular potentials to investigate the impact of a catheter electrode on the extracellular potential (V_e). Images show the difference in V_e at the shown time step based on a simulation including the catheter and a reference simulation. Scales at both axes indicate dimensions in mm. Shown figures are zoomed in to the catheter tip.

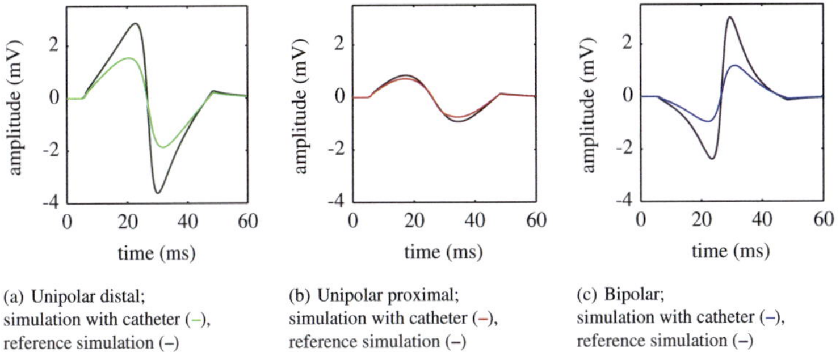

(a) Unipolar distal; simulation with catheter (–), reference simulation (–)

(b) Unipolar proximal; simulation with catheter (–), reference simulation (–)

(c) Bipolar; simulation with catheter (–), reference simulation (–)

Figure 12.3. Comparison of extracellular signals obtained from a simulation with an 8F ablation catheter and a reference simulation

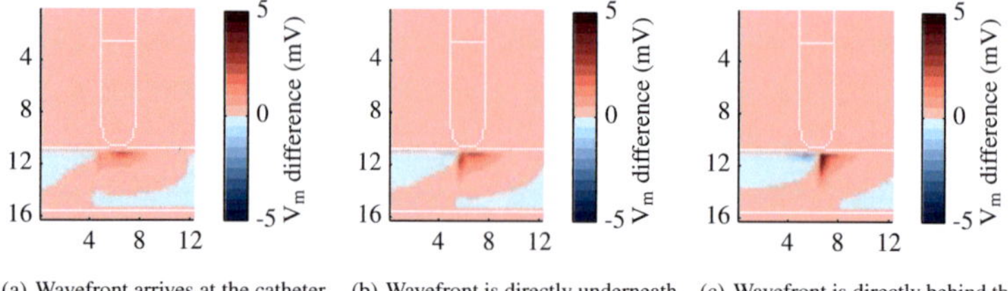

(a) Wavefront arrives at the catheter tip ($t = 25.5\,ms$)

(b) Wavefront is directly underneath the catheter ($t = 26.5\,ms$)

(c) Wavefront is directly behind the catheter ($t = 27.5\,ms$)

Figure 12.4. Time series of the difference in transmembrane voltage to investigate the impact of a catheter electrode on the transmembrane voltage (V_m). Images show the difference in V_m at the shown time step based on a simulation including the catheter and a reference simulation. Scales at both axes indicate dimensions in mm. Shown figures are zoomed in to the catheter tip.

Simulated transmembrane voltage was also affected by the presence of a catheter (Fig. 12.2). The voltage underneath the catheter was more positive, when the wavefront entered the area adjacent to the electrode. When the wavefront was underneath, also a negative difference to the reference simulation was observed. The maximum detected voltage difference was 6 mV, which indicates that all voxels were excited by the passing wavefront.

12.2.2 Influence of a Large Metal Surface on CV and Extracellular Current Distribution

Local activation times in the surface layer of the simulated patch were investigated to analyze the influence of a 2.3 mm catheter on conduction velocity. By visual comparison of the isochrone curvature, in the simulation with catheter, a slight conduction slowing can be observed before the wavefront enters the catheter area (Fig. 12.5). Underneath the catheter the distance between the isochrones was enlarged, which implies a higher conduction velocity. Local CV was 1.44 m/s compared to 1.2 m/s in the reference simulation.

Subsequently, a test setup with a 20 mm × 6 mm × 1 mm plate electrode was used (Fig. 12.1) in order to emphasize the effects of a metal electrode on CV. The resulting LATs are depicted in Fig. 12.6. Analogous to the previ-

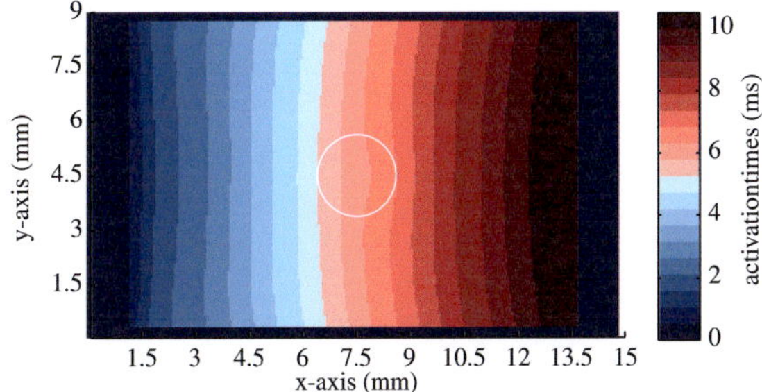

(a) Local activation times in the most endocardial layer of a simulated patch. The white circle indicates the location of a 2.3 mm diameter catheter.

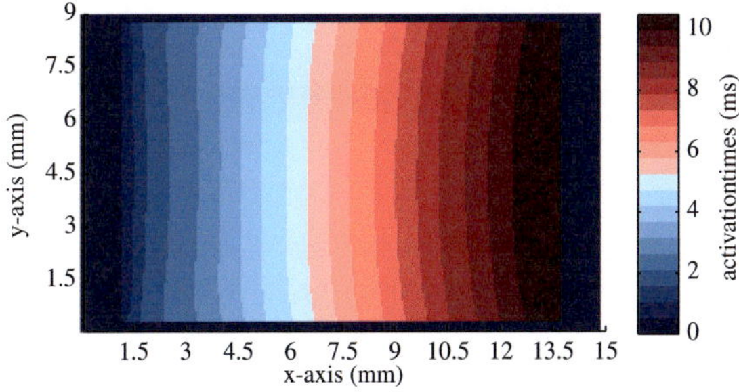

(b) Local activation times in the most endocardial layer of a simulated patch. This reference simulation was performed without a catheter.

Figure 12.5. Local activation times illustrating the impact of a 2.3 mm catheter on local conduction velocity

ous simulation, tissue underneath the electrode was activated earlier than in the adjacent areas. To visualize this effect, the LAT differences between the simulation including the plate electrode and the reference simulation were calculated (Fig 12.7). In the X-Y-plane view, positive difference values were extracted in the area in front of the electrode. In this area a conduction slowing was caused by the presence of the electrode. Underneath the electrode the difference switched to negative values, which stands for a higher CV underneath the electrode. At the end of the electrode the difference reached values above 1 ms. Directly behind the electrode a conduction slowing was detected. The X-Z-plane view shows, that conduction velocity was mainly

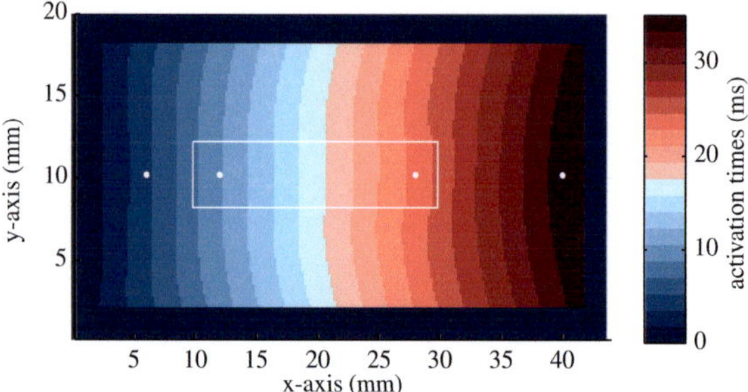

(a) Local activation times in the most endocardial layer of a simulated patch. Electrode location is indicated by a white rectangle; White dots indicate the measurement positions, which were used to extract conduction velocities shown in tab.12.1. Positions are enumerated from left to rigth.

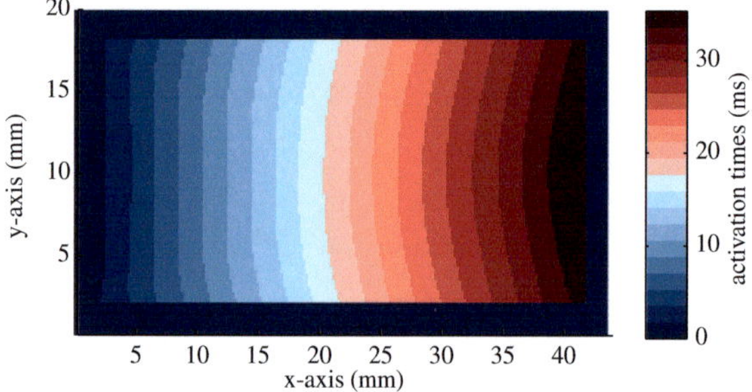

(b) Local activation times in the most endocardial layer of a reference simulation without electrode

Figure 12.6. Local activation times illustrating the impact of a $20\,\text{mm}\times6\,\text{mm}\times1\,\text{mm}$ plate electrode on local conduction velocity

influenced in the area directly underneath the electrode, which passed on the early activation to deeper layers. A quantitative analysis of CV development in the presence of a plate electrode is given in Tab. 12.1. Local CVs between the locations are indicated in Fig. 12.6. In the area in front of the electrode CV was reduced. Underneath the electrode CV was increased by 7.6 %. And in the area behind the electrode again conduction slowing was observed (-4.6 %).

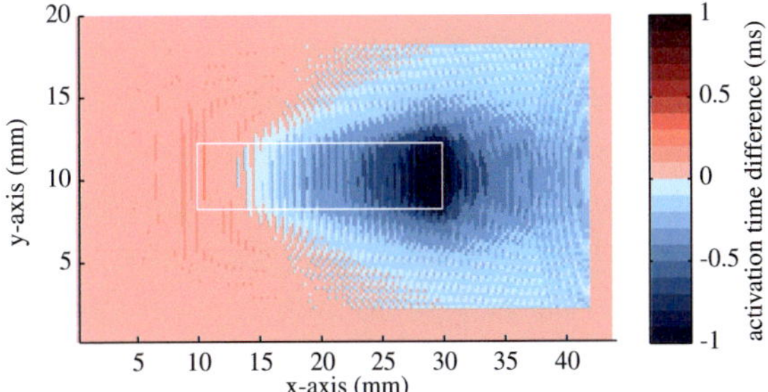

(a) X-Y-plane view, electrode area is indicated by white rectangle

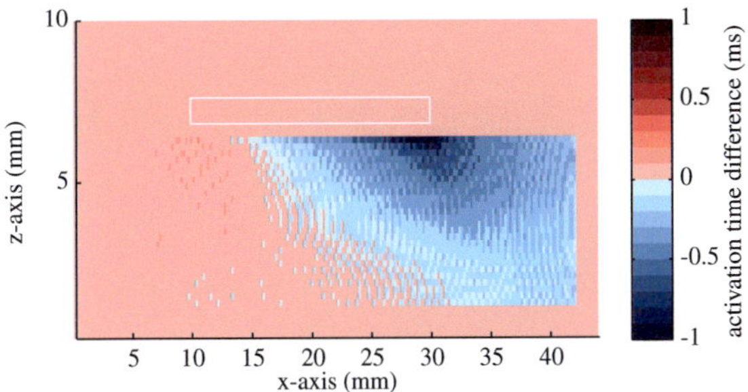

(b) X-Z-plane view, electrode area is indicated by white rectangle

Figure 12.7. Activation time differences illustrating the impact of a plate electrode on local conduction velocity

Table 12.1. Comparison of conduction velocities for the plate electrode simulation shown in figure 12.6. Positions were chosen to evaluate the influence of the electrode on CV. Position 1 is located before the wavefront enters the electrode area. Positions 2 and 3 are underneath the electrode. Position 4 is at the end of the simulated patch (see Fig. 12.6)

	CV position 1 to 2 (m/s)	CV position 2 to 3 (m/s)	CV position 3 to 4 (m/s)	CV position 1 to 4 (m/s)
With electrode	1.09	1.22	1.11	1.16
Without electrode	1.11	1.13	1.16	1.14
Relative difference (%)	-1.8	+7.6	-4.6	+1.3

As an additional parameter, which is influenced by the presence of an electrode, the extracellular currents were studied. This study is illustrated using a time series, covering the time, when the wavefront is passing the electrode (Fig. 12.8). Before the wavefront enters the electrode area a current flow along the electrode can be observed. Current flow along the electrode is increased when the wavefront is entirely underneath the electrode. At the point, where the wavefront starts to exit the electrode area, a smaller backward current flow can be observed, which keeps decreasing, till the wavefront has left the electrode area.

12.3 Discussion

By means of simulations this section addressed the question on the impact of a spatially extended electrode on electrophysiological properties of intra- and extracellular space. Four aspects were regarded. On the one hand, changes in transmembrane voltage and extracellular potentials were investigated. On the other hand conduction velocity and extracellular current flow in the electrode area were studied. The major effect of a catheter electrode on extracellular potentials was found in spatial averaging. The equipotential surface, formed by the catheter electrode produced a decrease in V_e variations during excitation propagation. This was also represented in reduced simulated EGM amplitudes in the presence of a large measurement electrode.

Small variations of transmembrane voltage were found. Underneath the electrode V_m was altered by up to 6 mV. This implies, that despite the electrical load imposed by the catheter electrode, the entire myocardium was activated. Otherwise differences would have been larger than the threshold voltage of -60 mV.

Conduction velocity underneath the simulated electrodes was increased. This effect can be explained by bath loading effects [161]. The increase in

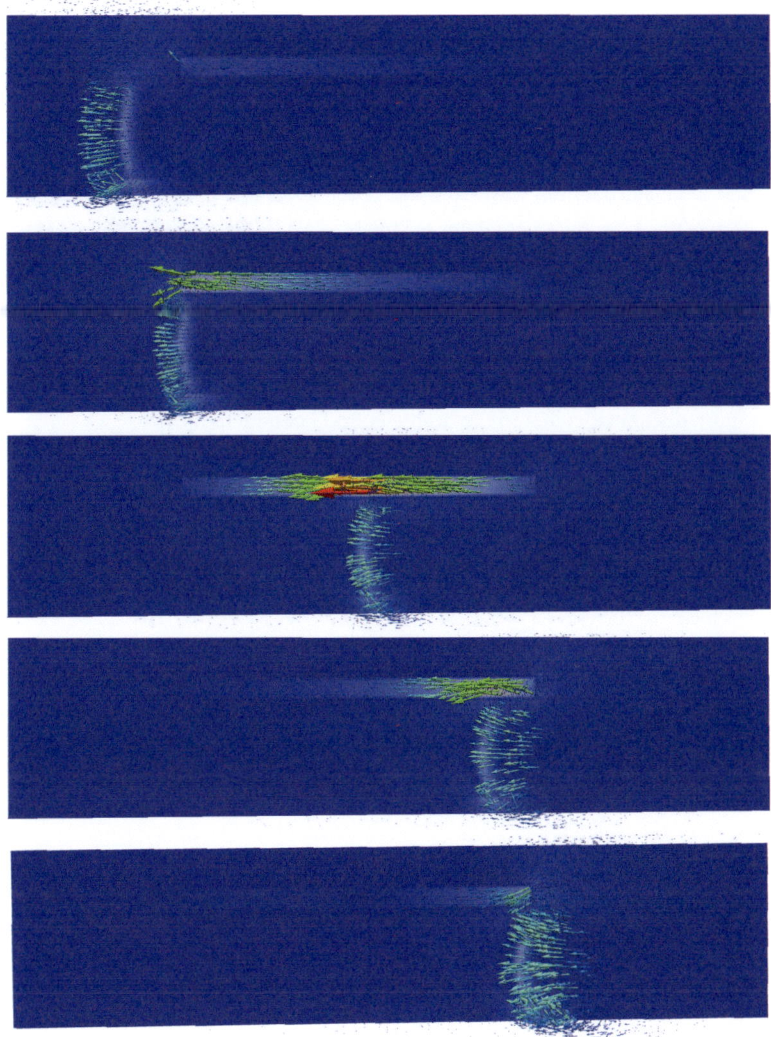

Figure 12.8. Extracellular currents for a wavefront passing underneath a plate electrode

CV of up to 10% is in agreement with a simulation study investigating the effect of multi-electrode arrays in myocardial propagation, which reported similar values in the case of densely pact electrodes [162]. Behind the electrode conduction slowing was observed. This effect can be related to a source sink mismatch due to wavefront curvature [163]. Due to the faster CV underneath the electrode, the wavefront had a convex shape behind the electrode, which caused the stated conduction slowing. An unexpected finding was the conduction slowing in front of the electrode. A possible explanation may be given by extracellular current flow. Shortly before the wavefront arrived at the electrode, a high current flow along the electrode was found. This may lead to a source sink mismatch due to the virtual connection of the wavefront to more distant cells due to the high electrode conductivity.

The results in this chapter showed, that highly conductive electrodes do have an effect on electrophysiological parameters. Especially in the case of large ablation electrodes and densely packed multi-electrode arrays this may affect measured extracellular potentials and conduction velocity.

13

Intracardiac Signals Measured on a Fibrotic Substrate

Fibrotic substrate is associated with areas of slow conduction, which are assumed to promote AF. Clinicians try to identify fibrotic areas by searching for complex fractionated atrial electrograms (CFAE). However, scientific discussion is still ongoing, if fibrotic substrate does cause electrogram fractionation. In this chapter, simulation results on fibrotic substrates are presented. Simulated catheter electrodes are resembling the dimensions of ring electrodes of conventional mapping catheters. Different electrode spacings are regarded (1-5 mm) by studying two different ring electrode catheters (*CathRing2mm*, *CathRing1mm*, Fig. 9.2). Two different types of fibrotic patterns were regarded (patchy and diffuse).

13.1 Applied Methods

Simulation Setups

In a planar patch of myocardium (l=11 mm, w=6.5 mm, h=0.65 mm) with a spatial resolution of 50 μm, fibrotic patterns were included using the method introduced in section. 9.1.3. Fibrotic volume fractions (FVF) were increased from 0 % to 50 % by steps of 10 %, which is in the range of clinical data [164]. Two types of mapping catheters (section 9.1.1) were placed onto the myocardium (Fig. 13.1). Using these catheter models bipolar signals with electrode distances of 1-5 mm could be studied. Due to the large number of voxels it was necessary to use the monodomain model to simulate cardiac

Figure 13.1. Ring electrodes *CathRing2mm* and *CathRing1mm* resembling clinical mapping catheter on a three dimensional patch of fibrotic myocardium

excitation. Extracellular potentials were restored by a forward calculation as described in (Sec. 6.2.2).

Conduction Velocity

For the two different fibrosis patterns and increasing fibrotic volume fraction global conduction velocities were analyzed. Therefore, local activation times at the start and at the end of the patch were determined from the maximum upstroke of V_m.

Comparison to a Clinical CFAE Database

Bipolar simulated electrograms were filtered with the clinical standard filter (Sec. 8.3). Subsequently, characteristic parameters were extracted, which were the number of positive peaks and the number of zero-crossings per EGM. Furthermore, the peak to peak amplitude and the mean time between positive peaks and between zero crossings was regarded. Results of this analysis were compared to mean values of the clinical CFAE database (Sec. 4.4.1). Characteristic features of simulated EGMs were compared to values extracted from a clinical CFAE database for each of the four CFAE classes.

13.2 Simulated EGMs on Diffuse Fibrosis

Diffuse fibrosis is characterized by a pattern of small dot like, homogeneously distributed islands of collagenous tissue [25]. With increasing fibrotic volume fraction the simulated EGMs display a drastic reduction in peak-to-peak amplitude (Fig. 13.2). Furthermore, EGM width is increased, which implies a reduced conduction velocity. Signals for high levels of fibrosis (40-50 %) show a slightly serrated flow, however no fractionation is visible. When regarding the filtered version of the EGMs, amplitude symme-

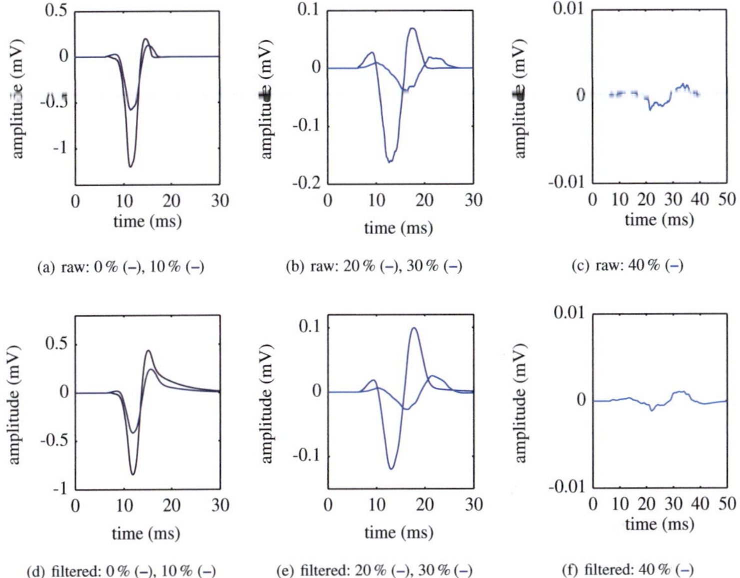

(a) raw: 0 % (–), 10 % (–) (b) raw: 20 % (–), 30 % (–) (c) raw: 40 % (–)

(d) filtered: 0 % (–), 10 % (–) (e) filtered: 20 % (–), 30 % (–) (f) filtered: 40 % (–)

Figure 13.2. Bipolar EGMs for different volume fractions of diffuse fibrosis (%); (a)-(c): raw signals; (d)-(e): EGMs filtered with a bandpass of 30-250 Hz

try is restored and the serrated curve shape is smoothened. Even though the EGM peak-to-peak amplitude is almost zero for 40 % fibrosis, conduction was still present. For 50 % of diffuse fibrosis conduction block occurred.

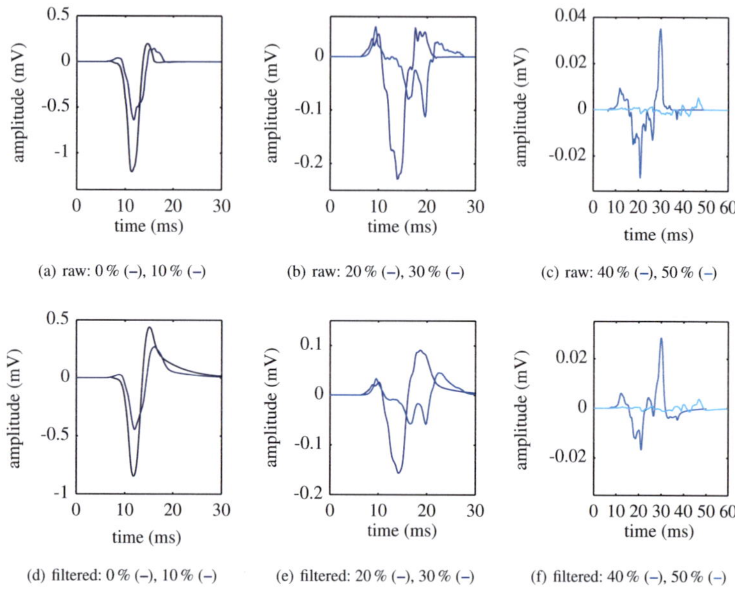

Figure 13.3. Bipolar EGMs for different volume fractions of patchy fibrosis (%); (a)-(c): raw signals; (d)-(e): EGMs filtered with a Bandpass of 30-250 Hz

13.3 Simulated EGMs on Patchy Fibrosis

Strand like, larger fibrotic areas, which are mostly oriented along myocardial fiber direction were characterized as patchy fibrosis [25]. Simulated EGMs on a model of patchy fibrosis displayed a serrated curve shape for levels above 20 %, and beginning fractionation above 30 % (Fig. 13.3). Analogous to the diffuse pattern, EGM amplitude decreased with increasing percentage of fibrosis. In the case of patchy fibrosis, propagation was present up to 50 % of fibrosis, the bipolar EGM amplitude was more than 100 times smaller than under physiological conditions. After applying a clinical bandpass filter the signal course was smoothened, however, the major peaks were preserved.

Motivated by physicians, reporting a temporally unstable behavior of electrogram fractionation, the influence of different propagation directions rel-

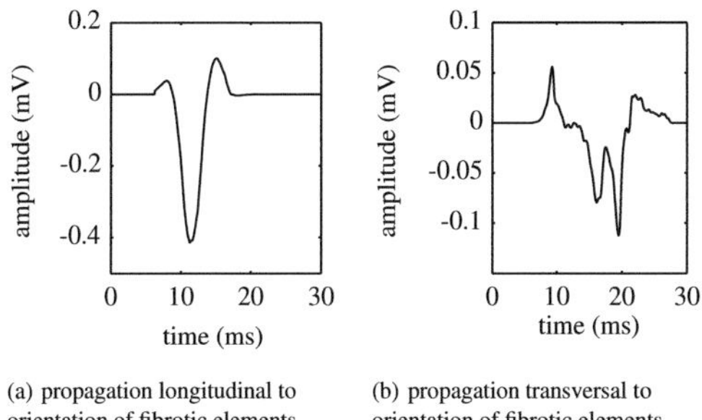

(a) propagation longitudinal to orientation of fibrotic elements

(b) propagation transversal to orientation of fibrotic elements

Figure 13.4. Bipolar EGM for a fibrotic volume fraction of 30 % (patchy fibrosis) for transversal and longitudinal propagation relative to the orientation of fibrotic elements

ative to the orientation of the fibrotic elements was analyzed. A simulation with 30 % fibrosis and propagation along the fibrotic element orientation produced a perfectly smooth signal. It was morphologically completely different from the signal with propagation transverse to the fibrotic elements (Fig. 13.4), which showed a high degree of fractionation.

The influence of different electrode spacings was investigated using a four electrode catheter with 1 mm electrode spacing (*CathRing1mm*, Fig 9.2(b)). Bipolar EGMs for 1 mm, 3 mm, 5 mm electrode spacing on simulated patchy fibrosis of 30 % and 40 % showed increased fractionation with increasing electrode distance (Fig. 13.5). While for an electrode spacing of 1 mm rather physiological EGMs were measured. EGMs for 5 mm spacing displayed pronounced fractionation. Standard clinical mapping catheters commonly provide an electrode spacing of 5 mm. Therefore, the latter simulations are well suited for comparison to the clinical database.

13.4 Conduction Velocity (CV)

In the presented simulations, with increasing fibrotic volume fraction the conduction velocity was reduced. Global conduction velocities, which were extracted from simulations for both regarded types of fibrosis showed an approximately linear decrease with increasing percentage of fibrosis (Fig. 13.6).

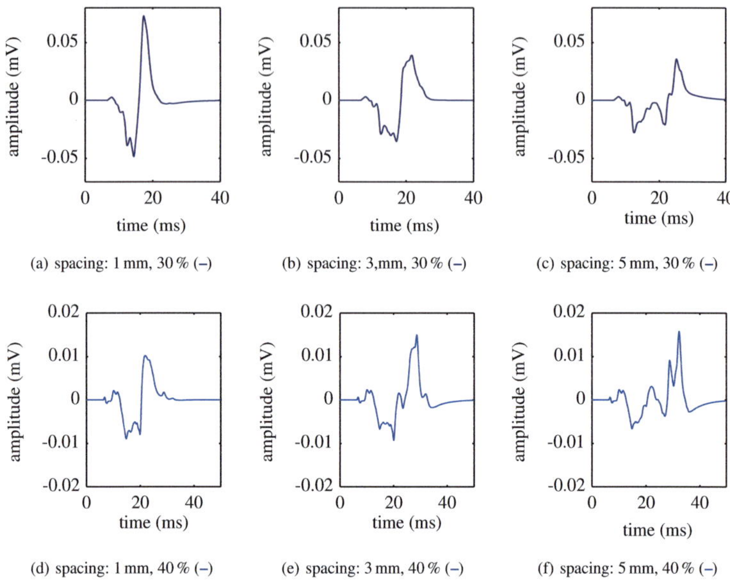

(a) spacing: 1 mm, 30 % (–)

(b) spacing: 3,mm, 30 % (–)

(c) spacing: 5 mm, 30 % (–)

(d) spacing: 1 mm, 40 % (–)

(e) spacing: 3 mm, 40 % (–)

(f) spacing: 5 mm, 40 % (–)

Figure 13.5. Bipolar electrograms from a multi ring electrode catheter (*CathRing1mm*) for fibrotic volume fractions of 30 % and 40 %. EGMs are a result from the same simulation with different electrode configurations: Electrodes 1 and 2: 1 mm spacing : Electrodes 1 and 3: 3 mm spacing, Electrodes 1 and 4: 5 mm spacing

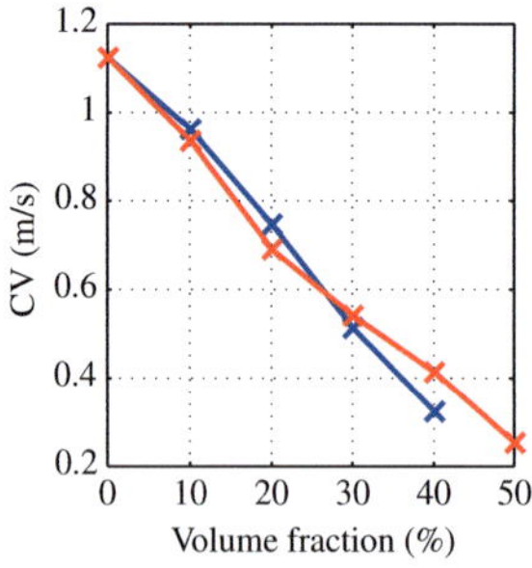

Figure 13.6. Development of global conduction velocity in simulations of fibrotic substrates. Patchy fibrosis (blue curve), diffuse fibrosis (red curve)

For the performed simulations the global conduction velocity turned out to be nearly independent from the fibrotic pattern. CV curves for patchy and diffuse fibrosis are close to each other with a maximum deviation of 0.1 m/s for a volume fraction of 40 %. However, for fibrosis of 50 % no CV could be estimated for the diffuse pattern, as conduction block occurred.

13.5 Comparison to a Database of Clinical Signals

The database of clinical CFAE signals, which was introduced in section 4.4.1 contains 605 classified electrograms, which were recorded after filtering with a bandpass filter (30-250 Hz). For comparison to clinical data, five temporal and morphological features were extracted from the clinical database, as well as from the simulated EGMs. These features were the peak-to-peak amplitude, the number of maxima, the mean time between maxima, the number of zero crossings and the mean time between zero-crossings within one EGM. The results of this analysis are displayed in table 13.1. Simulated EGMs with an electrode spacing of 5 mm are best suited for comparison to the measured data.

When comparing absolute amplitude values the simulated values differ by a factor of five from the measured values. This fact may result from the simulation method and will be addressed in the discussion. Relative amplitude changes from non fractionated to fractionated electrograms are 84 % for the clinical data (CFAE 0 vs CFAE 1) compared to 93 % for the simulation (0 % fibrosis vs 30 % fibrosis). Number of maxima and zero-crossings of simulated data (40 % and 50 %) are in the range of 5-10 per EGM. These numbers correspond with the clinical classes CFAE 1 and CFAE 2. Regarding the time between these events, the simulated values are smaller by more than a factor of 2.

13.6 Discussion

Several simulation studies have investigated potential mechanisms leading to electrogram fractionation due to fibrotic substrate. In [28, 112] a microstructural 2D model was presented. By introducing uncoupling between cells

Table 13.1. Characteristic features extracted from simulated and measured CFAEs. Displayed are the peak-to-peak amplitude (mV), mean number of maxima per active segment, mean time between maxima (ms), number of zero-crossings, mean time between zerocrossings (ms), for various electrode spacings **ES** and volume fractions of fibrosis **Fibr. (%)**.

ES (mm) / Fibr. (%)	peak-to-peak ampl. (mV)	# maxima	time (ms) b. maxima	# zero-cross.	time (ms) b. zero-cross.
1 / 0	1.07	2.00	7.00	2.00	5.00
1 / 10	0.53	2.00	6.00	4.00	4.67
1 / 20	0.31	2.00	8.00	4.00	4.67
1 / 30	0.12	2.00	9.00	4.00	5.00
1 / 40	0.02	6.00	5.20	5.00	5.75
1 / 50	0.01	7.00	7.17	8.00	4.43
2 / 0	1.39	2.00	8.00	2.00	6.00
2 / 10	0.78	2.00	7.00	2.00	5.00
2 / 20	0.27	2.00	9.00	2.00	6.00
2 / 30	0.11	3.00	6.50	3.00	7.00
2 / 40	0.05	5.00	6.00	6.00	4.80
2 / 50	0.00	9.00	4.25	11.00	3.40
3 / 0	1.22	2.00	8.00	2.00	6.00
3 / 10	0.51	2.00	9.00	4.00	4.67
3 / 20	0.30	2.00	8.00	4.00	5.00
3 / 30	0.08	5.00	3.25	4.00	6.67
3 / 40	0.02	7.00	4.33	7.00	4.33
3 / 50	0.01	8.00	5.29	8.00	5.86
5 / 0	0.92	2.00	10.00	2.00	8.00
5 / 10	0.40	3.00	5.50	3.00	5.00
5 / 20	0.19	3.00	5.50	3.00	6.00
5 / 30	0.06	4.00	5.67	3.00	8.00
5 / 40	0.02	5.00	6.50	7.00	4.67
5 / 50	0.01	8.00	5.86	10.00	4.67
Clinical CFAE database (electrode spacing 5mm)					
CFAE 0	5.53	3.67	12.73	4.60	10.63
CFAE 1	0.86	4.84	14.55	6.21	11.54
CFAE 2	0.78	8.14	14.25	10.45	11.21
CFAE 3	0.63	21.04	14.90	26.07	11.96

electrogram fractionation was simulated. The influence of the diameter of a disk electrode (up to 0.96 mm) and a simple bipolar configuration (0.96 mm electrodes with a distance of 0.23 mm) was investigated [28]. In [116] a quasi 3D (two layer) model of fibrosis was created, based on histological data of the rabbit isthmus. In this study it was shown, that a microelectrode based sensor delivers different signal characteristics from different types of fibrosis. These features were used to automatically distinguish between the different patterns. Further simulation studies addressing the impact of fibrosis on cardiac excitation patterns were presented in [165, 166], where approaches for patient specific modeling of fibrosis were presented. An efficient approach to incorporated fibrotic clefts by uncoupling of nodes was introduced in [167]

The focus of the study presented in this work, was electrogram fractionation in a clinical context. As the myocardial volume contributing to intracardiac EGMs reaches up to several millimeters from the electrode, a small three-dimensional setup was created. Furthermore, ring electrodes with a high conductivity where introduced. Using this approach the effect of electrode spacing in bipolar electrograms could be investigated on the same scale, which is used in clinical mapping catheters (electrode spacing 1-5 mm). The analysis of the simulations focused on bipolar EGMs, because CFAEs in their clinical application are only defined for this signal type [5]. Due to the small simulation setup, the position of a reference electrode for unipolar EGMs would be close to the electrodes. Therefore a UEGM would rather reflect a quasi bipolar signal without significance. Finally, characteristic features of simulated EGMs were directly compared to a database of clinically acquired CFAE signals.

Major findings of this study were a dependence of fractionation seen in bipolar electrograms on the type of fibrosis. Whereas diffuse fibrosis produced normally shaped EGMs with a slightly serrated flow, patchy fibrosis caused fractionation for a percentage above 30 %. According to the study presented in [116], a microelectrode catheter might be capable to detect micro fibrotic structures. Also, the propagation direction relative to the orientation of fibrotic elements influenced the occurrence of EGM fractionation. If the wavefront propagated along the fibrotic element orientation, which usually corresponds with the fiber direction, no fractionation was present in simulated EGMs. Another parameter which influenced the degree of fractionation was found in the electrode distance. Bipolar EGMs for 5 mm electrode spacing, showed increased fractionation compared to an electrode spacing of 1 mm.

Considering the fact, that not EGM fractionation it self, but the slow conduction is a potential mechanism behind AF, areas displaying diffuse fibrosis may have the same hazardous potential as areas with patchy fibrosis. This assumption is motivated by the finding, that regardless of the pattern of fibrosis CV decrease did only depend on fibrotic volume fraction with an approximately linear relationship to fibrotic volume. For the case of diffuse fibrosis the same results were found in [29]. Furthermore, it was found, that highly fibrotic areas are still a pathway for slow conduction. However, the simulated EGM amplitudes were small and may remain unnoticed under clinical measurement conditions.

Creation of fibrotic patterns was based on a statistical approach and therefore produced idealized results. This did not limit the goals of the study, which were to study the influence of known fibrotic patterns on electrodes resembling catheters in clinical use. Furthermore, simulations were performed in a two step process. A monodomain simulation was used to calculate the propagation within the myocardium. Subsequently, extracellular potentials were forward calculated based on transmembrane currents. This uncoupled process may influence the significance of the absolute amplitude of simulated EGM. The impact on EGM morphology should be minor as the shape of the wavefront in this special case was mainly affected by fibrotic patterns and not by bath loading effects.

Simulated EGMs were directly compared to a database of clinical CFAEs. Analogous to clinical data a drastic reduction of amplitude was observed between fractionated and non-fractionated EGMs. This effect seems to be related to the reduced number of sources and the serrated wavefront in fibrotic areas. Also the number of deflections and zero-crossings was in good agreement with clinical classes CFAE 1 and CFAE 2.

Temporal behavior, reflected by the mean time between zero-crossings was not fully met. The difference between measured and simulated data was about 5-7 ms and therefore smaller as in the clinical data (12-15 ms). This might be related to the high local conductivity in the simulations, which would slow down, when taking electrophysiological remodeling into account [28]. Apart from the stated small deviation, the reflected temporal characterization may still be meaningful. Among the possible wavefront patterns, which are discussed to cause electrogram fractionation, the temporal behavior of EGMs in this study is closer to the clinical data than e.g. the rotor pattern. In this case cycle lengths around 200 ms would be anticipated [145].

14

Intracardiac Electrograms Measured on Acute Ablation Lesions

Controlling the scar formation during RFA is an open task to optimize the therapy of cardiac arrhythmias. Alterations in IEGM morphology have been shown to be an important contributing candidate in this context [6, 168]. Although experimental and clinical studies have observed characteristic EGM changes during ablation, the underlying pathological mechanisms are barely understood. In this study, a simulation environment for intracardiac EGMs on acute ablation lesions was developed. The simulation environment was parametrized by literature data introduced in chapter 3.3. As literature data on electrophysiological changes during hyperthermia is sparse, an experimental study investigating CV development under hyperthermic conditions was conducted. Using the developed lesion model, simulated signals were directly compared to clinically measured EGMs.

14.1 Applied Methods

Experimental Investigation of Conduction Velocity Under Hyperthermic Conditions

In five right atrial preparations of adult male Fisher rats (F344) with a weight of 455 ± 10 g temperature profiles were acquired following the protocol described in Sec. 7.5.5. During acquisition, preparations were paced to override sinus rhythm. Pacing frequency was set to 5 Hz in four of the five preparations. In one preparation it had to be increased to 5.5 Hz due to a higher sinus rhythm frequency. Temperature was increased until pace capture was lost due to rising sinus rhythm frequency caused by elevated temperature. In all preparations a level of 42 °C was reached. In two preparations additional measurements at 43.8 °C were acquired. Based on optically recorded propagation patterns and anatomical characteristics, regions of interest were chosen for CV estimation (Fig. 14.1). This resulted in ten different regions of interest. Within these regions, CV was obtained from the barycenters of LAT areas as described in Sec. 7.2.2. For each temperature step the mean CV was calculated. In order to compare temperature development of different preparations, CV was normalized to baseline CV at 36.7 °C. If more than one measurement was available for this temperature, the mean of these measurements was taken for normalization.

Simulation Setups

IEGMs were simulated for orthogonal and parallel catheter orientation (see Fig.14.2) using a model of an 8F ablation catheter with an 8 mm tip electrode (Sec. 9.1.1). Tissue dimensions were 24 mm×24 mm×3 mm for orthogonal orientation and 32 mm×24 mm×3 mm for the parallel case. The catheter tip for orthogonal catheter orientation, was assumed to deform the tissue surface leading to an indentation depth of 1 mm. In the parallel case the catheter electrodes were fully in contact with the myocardium without deforming the surface.

Using the introduced model for acute ablation lesions (Sec. 9.1.4), a series of eight lesion stages up to complete transmurality was created. The lesion depths were 0, 7, 28, 50, 71, 92, 100 and 110% of the myocardial patch thickness (h). A depth of 110% represented a lesion, which was completely transmural and had a width increased by 10% compared to the 100% case. Depth to width ratio was set to 1.25 for orthogonal and 1.6 for parallel catheter orientation resembling the data shown in [6].

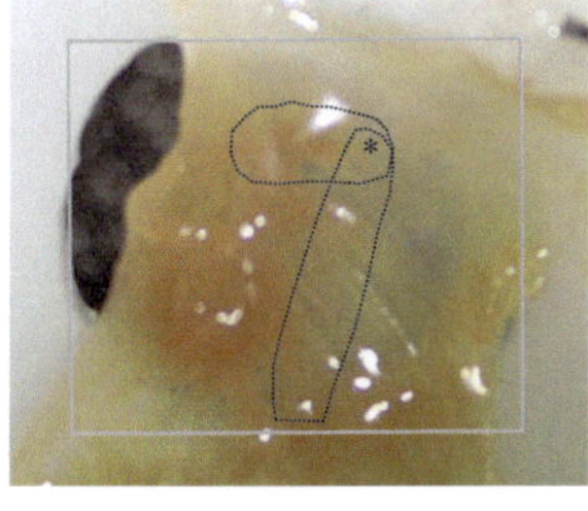

(a) Photographic image of the endocardial view of an atrial preparation. An asterisk (*) indicates the position of the stimulus electrode. The field of view of the optical system is delineated by a gray rectangle. Areas of CV estimation are encircled by dotted lines.

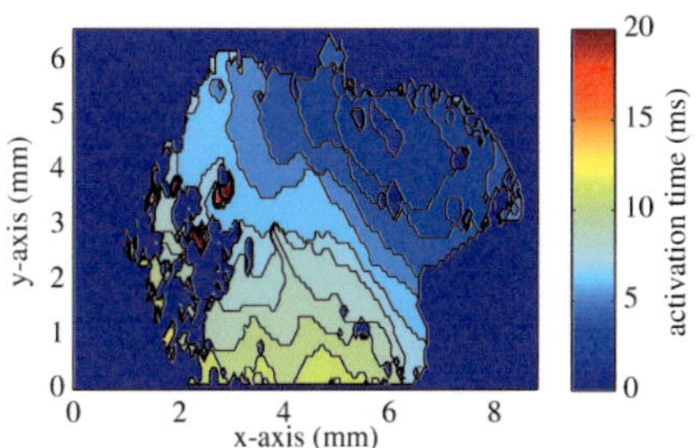

(b) Activation time map obtained from optical mapping data. Activation starts at the position of the stimulus electrode indicated in (a).

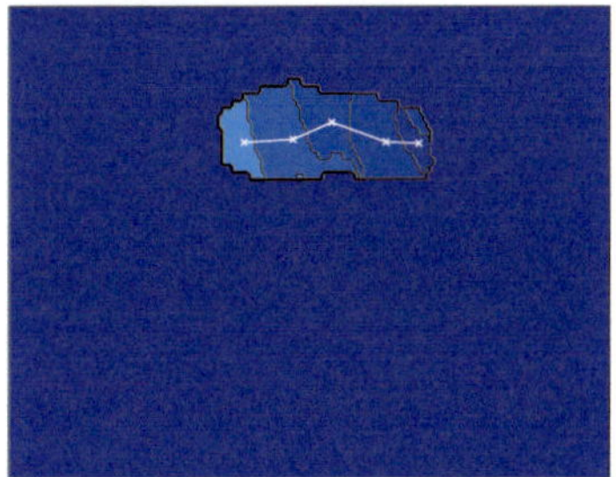

(c) CV estimation path in region 1. Barycenters of activation time areas are connected by a white line. Dimensions and colormap are same as in (b).

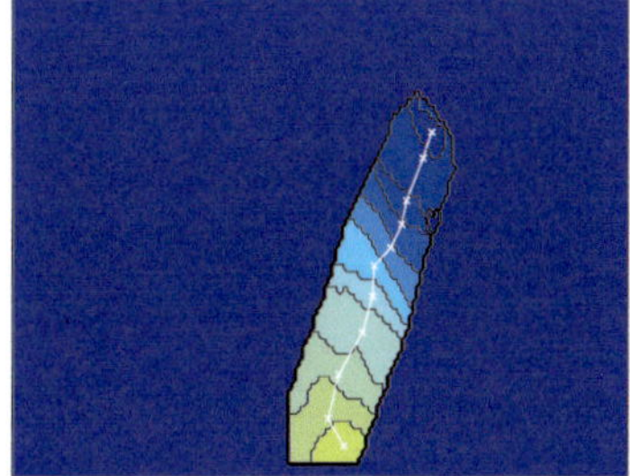

(d) CV estimation path in region 2. Barycenters of activation time areas are connected by a white line. Dimensions and colormap are same as in (b).

Figure 14.1. Visualization of exemplary regions used for CV estimation.

The lesion model was applied in different degrees of complexity in order to gain a more detailed understanding of the influence of the components. Therefore, the lesion model was divided into three components:

- Passive necrotic lesion core
- Lesion border zone with changed AP characteristics
- Lesion border zone with changed CV characteristics

Additional simulations were performed, in which a passive layer of connective tissue was added at the endo- and epicardial surface. The influence of

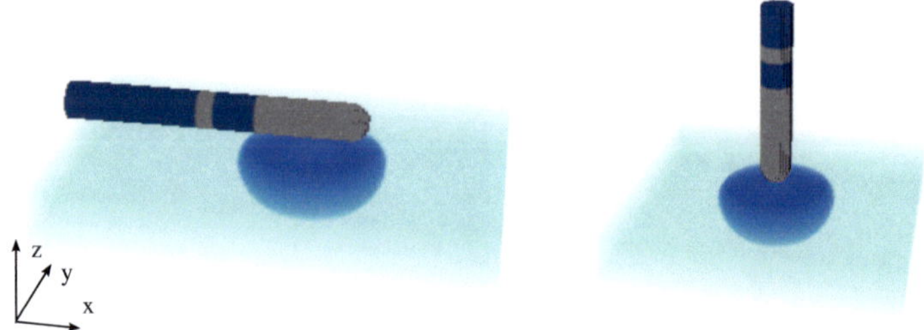

Figure 14.2. Model of an 8F ablation catheter in a simulation setup containing a planar patch of myocardium (light blue) and an ellipsoidal shaped lesion (dark blue). The catheter is oriented parallel (left) and orthogonal (right) to the tissue surface.

each of the components on EGM morphology was investigated in various simulations. The impact of different width to depth ratios (1.25, 1.4, 1.6 1.8) was assessed for a fixed lesion depth of 92% and orthogonal catheter orientation.

Comparison of Filtered Simulated EGM to Clinical Signals

11 datasets recorded in five patients were compared to simulated data. Each dataset contained EGMs recorded while point lesions were created by RFA. Signals were acquired at least 10 s before ablation. Ablation sequences lasted between 20 s and 60 s, followed by at least 10 s of recording after the end of RF application. For a detailed description of the data acquisition and signals in the datasets, see Sec. 8.1. Catheter orientations were identified by fluoroscopy and classified in parallel and non-parallel orientation relative to the tissue surface. Unipolar and bipolar electrogram templates over signal pieces of 5 s were created by an automatic segmentation algorithm (Sec. 8.2).

Clinical signal templates were quantitatively compared to simulated data. Simulated signals were down sampled to 1 kHz in order to match the sampling rate of clinical EGM for quantitative comparison. Moreover, the clinical filter function (Bandpass 30-250 Hz, Sec. 8.3) was applied to the simulated signals. Relative changes of fiducial markers before and after ablation were calculated. These were (1) the maximum positive peak amplitude, (2) the maximum negative peak amplitude (3) the maximum deflection between these points (see also Sec. 9.2.2). Amplitude development over time was analyzed by creating templates of 1 s windows for three representative clinical datasets.

Translation of Electrogram Parameters to Myocardial Excitation

In clinical application IEGMs are a mean to diagnose excitation patterns inside the myocardium. In order to translate characteristic signal points to myocardial excitation, fiducial points in the EGM were compared to the corresponding position of the excitation front determined from LATs.

14.2 Conduction Velocity Development under Hyperthermic Conditions

Data of an exemplary temperature profile shows an increase in CV between 36.7 °C and 42.0 °C (Fig. 14.3(a)). Baseline CV at 36.7 °C was 0.69 m/s, a maximum CV (0.92 m/s) was reached at 42.0 °C. When temperature was further increased to 43.8 °C, CV reduced to 0.89 m/s. After returning to baseline temperature, CV upon return was 0.77 m/s and after a 10 min waiting period 0.81 m/s was measured. Combined data of all ten datasets (Fig. 14.3(b), (c)) displays a similar course. Median values monotonically increase between 36.7 °C and 42.0 °C. Baseline CV at 36.7 °C was 0.64±0.13 ms. At 42.0 °C median CV was 119% of baseline CV. At 43.8 °C CV was decreased compared to the maximum value (median: 112%). After returning to baseline temperature, median CV remained elevated by 4% compared to the value before the temperature profile.

14.3 Lesion Characteristics

In chapter 3, electrophysiological changes in myocardial tissue during ablation were summarized. In order to reproduce these changes, conduction velocity and action potential characteristics in the border zone surrounding the necrotic lesion core were adjusted.

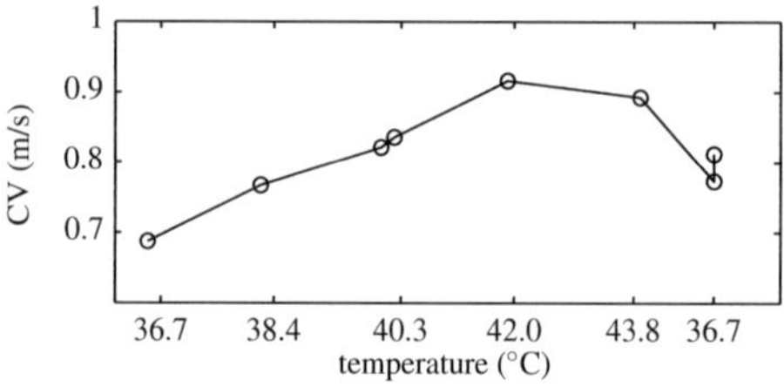

(a) Exemplary CV development for elevated temperatures

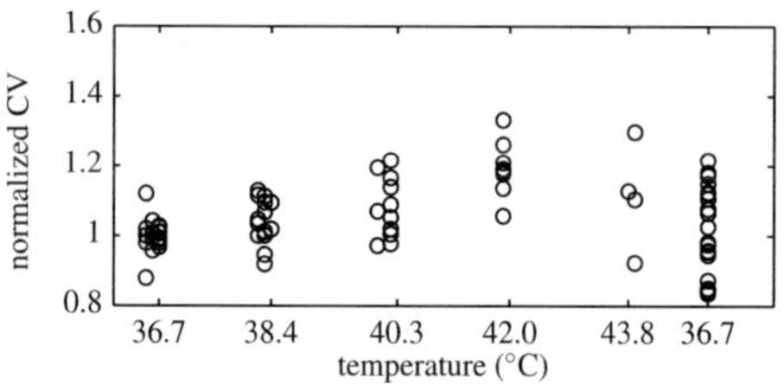

(b) Normalized CV values of five right atrial preparations relative to bath temperature

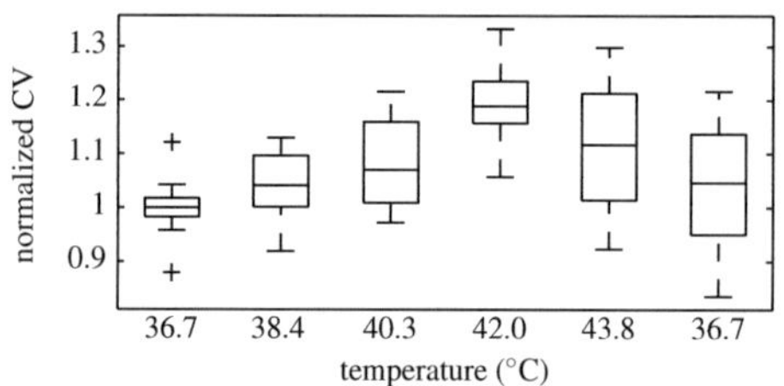

(c) Boxplot representation of the data shown in (b)

Figure 14.3. Experimentally obtained data on CV development under hyperthermic conditions of 10 regions of interest. Baseline CV at 36.7 °C was 0.64±0.13 m/s

14.3.1 AP Characteristics

In a first step, the model for ischemic ventricular cells [105] was adjusted to the reported AP changes. This was done on tissue level using small pieces of myocardium of $50 \times 10 \times 10$ voxels (spatial resolution 0.2 mm). It was found, that reported AP changes were macroscopically reproduced by cells adapted to ischemia phase 1a (Tab. 14.1). The resting membrane potential, which was

Table 14.1. Control values for action potential parameters derived from tissue level simulations and resulting changes in the scar border zone for a zone factor ZF=1 resulting from the ischemia model

	$V_{m,rest}$ (mV)	APA (mV)	dV/dt_{max} (V/s)	APD (ms)	CV (m/s)
Control	-85.48	56.49	107.46	310	0.83
Ischemia phase 1a	-73.48	26.72	87.45	177	0.5
Change %	+14	-52	-18	-42	-40

elevated by 14% and the action potential amplitude, which was reduced by 52% are most important for EGM formation. Due to the reduced AP upstroke velocity (dV/dt, -18%) also the conduction velocity was reduced by 40%. These reported values are valid for the innermost layer of the border zone, which represents temperatures above 49 °C. Moving away from the lesion core, AP changes were linearly decreased by weighting with a zone factor. The zone factor was 1 at the 49 °C iso-surface and 0 at the 43 °C iso-surface. Consequently, AP changes decreased linearly with the of zone factor down to 0 at the 43 °C iso-surface.

14.3.2 Conduction Velocity

Intracellular conductivity was adjusted in order to reproduce the CV profile reported in [50]. An initial guess was received by tissue simulations of small pieces of myocardium. However, conductivities had to be further reduced to achieve the respective CV values. Final conductivity parameters, which were used in the lesion area are listed in Tab. 14.2. Baseline conductivity of normal myocardium was 0.5 S/m. Conductivity was increased in the outer border zone and subsequently decreased closer to the lesion core. Resulting LATs and an estimate of the local conduction velocity are depicted in Fig. 14.4. In the area surrounding the lesion core, the isochrone curvature was increased compared to the other parts of the simulated myocardium. The outer border zone was activated earlier, whereas the inner part of the border zone was activated later than the surrounding tissue. Consequently, when approaching

Table 14.2. Resulting conductivity values after the adaptation of CV parameters in the scar border zone; extracellular conductivities σ_e, intracellular conductivities σ_i in S/m

	σ_i	σ_e
Lesion (necrotic)	0	0.1
Lesion (40 °C)	0.5	0.2
Lesion (41 °C)	1.0	0.2
Lesion (43 °C)	0.35	0.2
Lesion (45 °C)	0.1	0.2
Lesion (47 °C)	0.1	0.2
Lesion (49 °C)	0.05	0.2

the lesion core, local CV values first increased up to 1 m/s (+25%) and then gradually decreased down to 0.3 m/s (-63%) compared to baseline CV of 0.8 m/s.

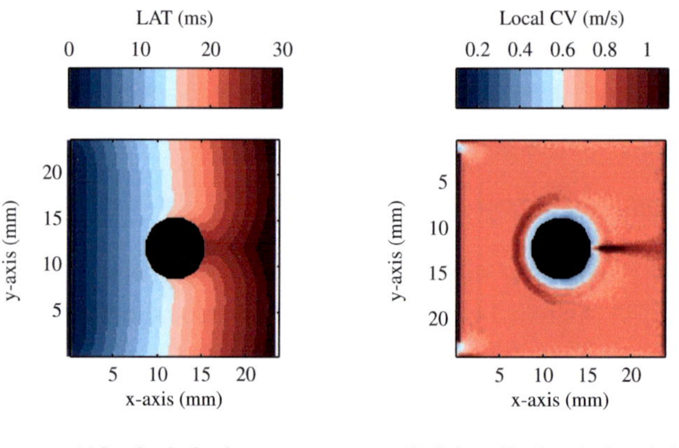

(a) Local activation times   (b) Estimated local conduction velocities

Figure 14.4. X-Y-plane view on activation times and estimated CV 1 mm below the endocardial surface of the simulated patch. The location of the necrotic core of the lesion is indicated by a black circle. In the vicinity of the lesion core isochrones were bent backwards, which reflects conduction slowing in the lesion border zone. The CV plot shows increased values in the wider border zone and a CV decrease surrounding the necrotic core.

14.4 Clinical Signals

Representative clinical recordings during RFA are presented in the following. EGMs recorded at one position before, during and after ablation are displayed in figures 14.5 and 14.6. EGM morphology did not alter over a timeframe of 20 s before ablation for orthogonal catheter orientation (Fig. 14.5).

During the ablation sequence the negative peak amplitude of the distal UEGM decreased to a value of 0 mV. In contrast to that, the positive morphology hardly changed. Also the downstroke velocity of the negative peak gradually decreased. Bipolar EGM for this catheter orientation was almost an inverted version of the distal UEGM. Therefore, the described changes can be translated by inversion of the polarity. Within 20 s after the ablation sequence, EGM morphology in all leads did not change further.

In the case of parallel catheter orientation, both electrodes are in contact with the tissue surface (Fig. 14.2). Therefore, both UEGMs present independent signals. Signal morphology was stable over a 5 s timeframe before ablation (Fig. 14.6). Both UEGMs showed a biphasic morphology. The mathematical difference of both UEGMs is represented in the bipolar EGM. In the presented case, this led to a biphasic morphology, where the distal UEGM contributes with inverted polarity. During the ablation sequence, the distal peak entirely vanished. In the proximal UEGM only a slight amplitude reduction is visible. The negative peak of the bipolar EGM, which is formed by the proximal EGM, remained almost unchanged, while the positive peak strongly decreased. After the ablation sequence, EGM morphology was again stable over 20 s. Major EGM changes occurred between 5-10 s after ablation onset, independent of the catheter orientation. This behavior can also be seen in the charts in Fig. 14.7, where the distal UEGM amplitude was evaluated during the ablation sequence. According to the shown EGMs, the changes in myocardial properties seem to be non-reversible after RFA.

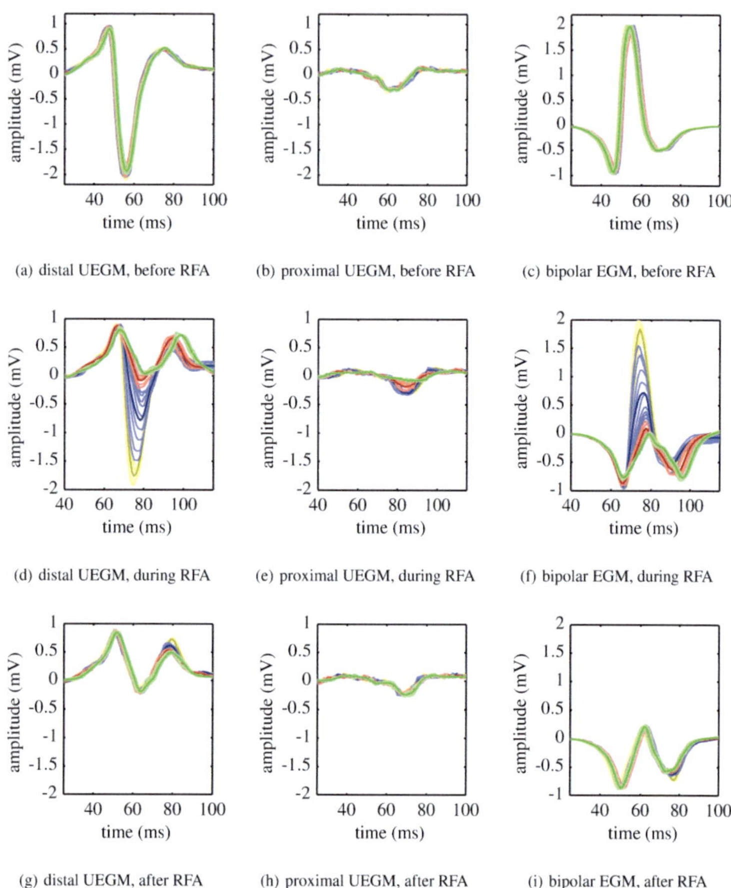

(a) distal UEGM, before RFA   (b) proximal UEGM, before RFA   (c) bipolar EGM, before RFA

(d) distal UEGM, during RFA   (e) proximal UEGM, during RFA   (f) bipolar EGM, during RFA

(g) distal UEGM, after RFA   (h) proximal UEGM, after RFA   (i) bipolar EGM, after RFA

Figure 14.5. Clinical signals for orthogonal catheter orientation. Top row: signals before RFA, mid row: signals during RFA, bottom row: signals after RFA. Color code indicates the temporal development within the 20 s recordings: Yellow: 0-5 s, blue: 5-10 s, red: 10-15 s, green:15-20 s; respective dark colors indicate the mean of all EGMs within the time frame.

A different behavior was observed in one example, which is shown in Fig. 14.8. Here, a series of 3 ablation intervals with a duration of 20 s each, was applied at one location. In between RFA applications breaks of 6 s and 7 s led to partially reversible EGM changes. The figure shows bipolar sig-

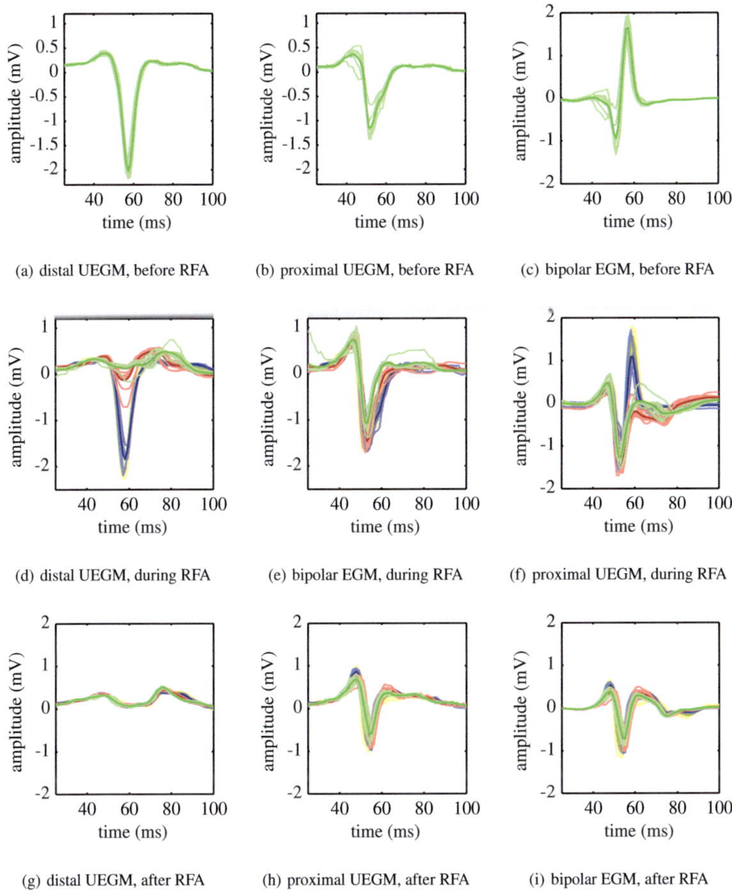

Figure 14.6. Clinical signals for parallel catheter orientation. Top row: signals before RFA, mid row: signals during RFA, bottom row: signals after RFA. Color code indicates the temporal development within the 20 s recordings: Yellow: 0-5 s, blue: 5-10 s, red: 10-15 s, green:15-20 s; Respective dark colors indicate the mean of all EGMs within the time frame.

nals for a parallel catheter orientation. Based on the electrogram shape and the ablation related changes, a propagation direction from the proximal towards the distal electrode can be assumed. During the first RFA application, bipolar positive peak amplitude decrease was most pronounced with a de-

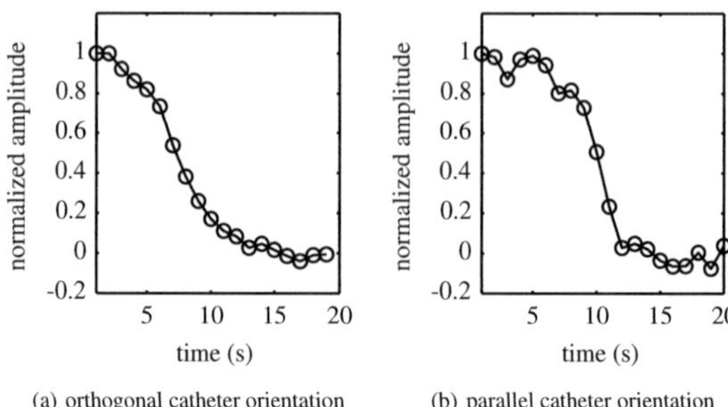

Figure 14.7. Temporal development of the negative peak amplitude of the distal UEGM for two example signals. Each data point represents a mean template over 1 s. Major changes in both cases occurred within the first 10 s of the RFA sequence.

crease from 2 mV down to 0 mV. In contrast to the examples shown before, the amplitude recovered to a certain extent during the RFA pause. Within the second RFA application a similar state compared to the first application was reached.

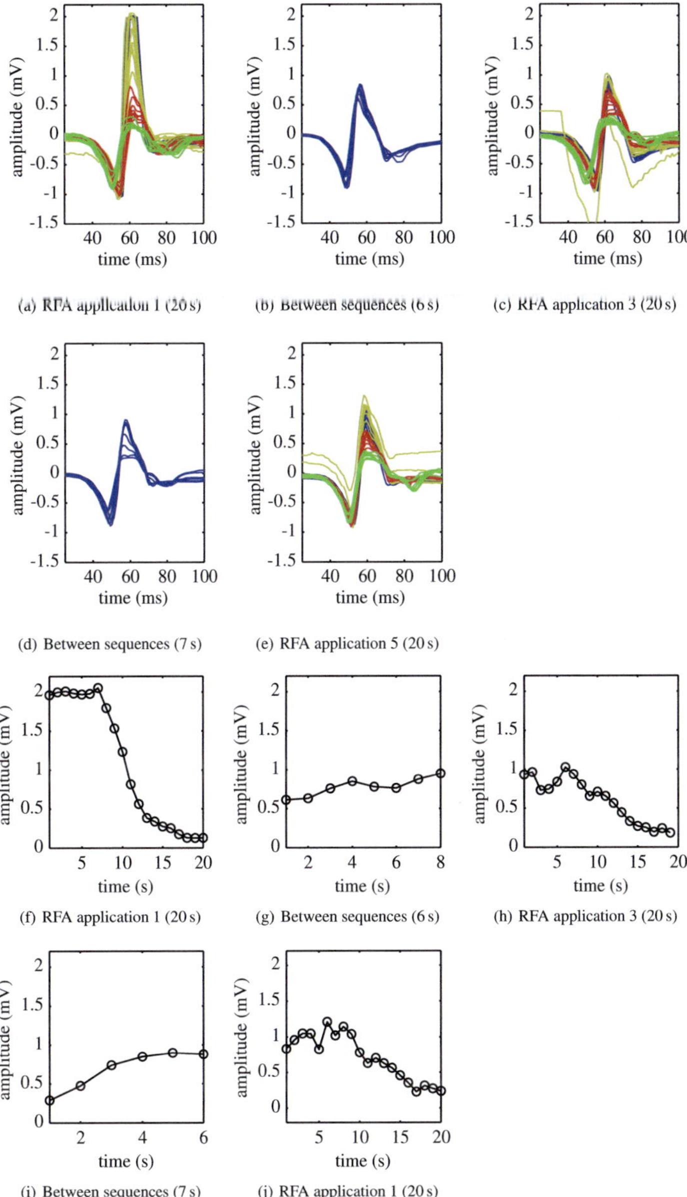

Figure 14.8. Exemplary bipolar recording sequence for one RFA target location. Signals are recorded during (a,c,e) and in between RFA applications (c,d). (f-j) show the temporal development of the positive peak amplitude. Different to the signals shown before is, that positive peak amplitude seems to 'recover' in between the RFA applications.

Again, when the RFA application was paused, EGM changes were reversible. EGMs from the third ablation sequence look nearly identical to those from the second application. In the second part of the figure also the amplitude development of the regarded EGMs is visualized. Here, the recovery of the EGM amplitude after switching RFA off can be seen. In between the end of the first ablation sequence (Fig. 14.8(f)) and the recording of the ablation pause (Fig. 14.8(g)), 2 s of signal are missing, which can explain the 'jump' in EGM amplitude between the two displayed charts. A remarkable aspect is that the amplitude at the end of the first ablation sequence (0.13 mV) was smallest. Amplitudes at the end of sequence 2 and 3 showed an increasing tendency (0.18 mV, 0.24 mV).

14.5 Simulated Signals

Clinical EGMs presented in the last section showed major changes in the negative peak amplitude of distal UEGMs. These signals were bandpass filtered by the described clinical recording system (30-250 Hz). Therefore, the first part of this section presents simulated EGMs, which were also filtered using the identified filter function (Sec. 10.3.5).

First, the influence of different components of the lesion model on EGM morphology is presented for a fixed lesion depth of 92%. This analysis was performed without incorporating passive endo- and epicardial surface layers. Subsequently, EGMs of two sets of simulations are first regarded morphologically, followed by a quantitative analysis of the changes of characteristic EGM features during the ablation process. In these two sets, the complete lesion model was applied, whereas one set also accounts for a passive endo- and epicardial layer at the tissue surface. The final part focuses on the changes and the identification of fiducial points in unfiltered simulated EGMs.

14.5.1 Stepwise Investigation of Different Aspects of the Lesion Model

In a first step, the lesion was simulated by accounting only for the necrotic core for a lesion depth of 92%. This produced a negative peak amplitude of -3 mV in the distal UEGM (Fig. 14.9). Compared to the case without lesion (Fig. 14.10), the amplitude is reduced by 3 mV. When adding AP related changes in the border zone, the amplitude further decreased. Additionally, maximum downstroke velocity was reduced. When CV related changes were applied, morphological alterations were similar, however peak amplitude was lowered by 0.3 mV. In comparison, amplitude and downstroke changes were more pronounced due to CV related changes. For the full lesion model, simulated UEGM amplitude and downstroke velocity were further decreased. Comparing the chronic lesion model only described by a necrotic core and the acute lesion model including a modified border zone, clear differences in peak amplitude and peakedness could be found.

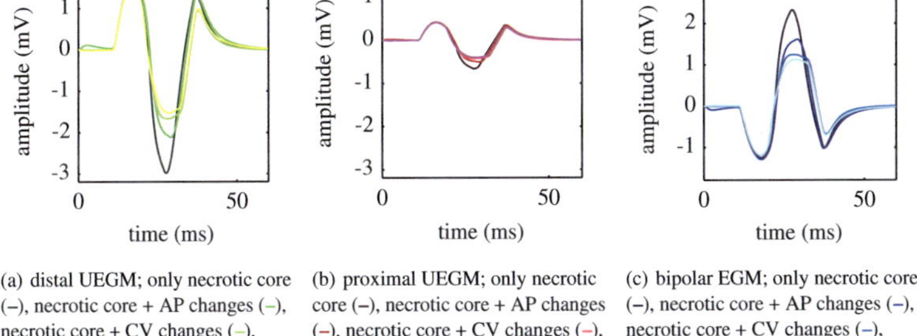

(a) distal UEGM; only necrotic core (–), necrotic core + AP changes (–), necrotic core + CV changes (–), complete lesion model (–)

(b) proximal UEGM; only necrotic core (–), necrotic core + AP changes (–), necrotic core + CV changes (–), complete lesion model (–)

(c) bipolar EGM; only necrotic core (–), necrotic core + AP changes (–), necrotic core + CV changes (–), complete lesion model (–)

Figure 14.9. Filtered simulated EGM for a lesion transmurality of 92% and orthogonal catheter orientation. In a stepwise approach, the complexity of the used lesion model was increased. Starting from a passive necrotic core, AP and CV effects were included in the borderzone. Finally, the complete lesion model with necrotic core and a border zone accounting for AP and CV changes was used.

Due to the orthogonal catheter orientation, proximal EGMs only displayed the farfield component of the distal EGM. Therefore, differences between simulated EGMs were small. Bipolar EGMs show similar changes as distal UEGMs but with inverted polarity.

14.5.2 Filtered Simulated EGMs at Different Stages of Lesion Transmurality

Electrogram changes due to increasing lesion size underneath the distal electrode of the ablation catheter were simulated. This process was studied regarding three catheter orientations. On the one hand, the catheter was oriented orthogonally to the tissue surface. On the other hand, two parallel catheter orientations were assessed. For parallel catheter orientation, propagation from the distal towards the proximal electrode was simulated, as well as the opposite direction. The full lesion model was applied (Fig. 14.10), while in one set of simulations, an additional passive layer of connective tissue at the endo- and epicardial surface was included (Fig. 14.11). It turned out, that this feature did not influence the qualitative outcome of the observed changes. However, amplitudes of simulated EGMs differed in the quantitative analysis. Therefore, qualitative changes will be described in a combined view.

Orthogonal Catheter Orientation

Analogous to the presented clinical data, changes in the distal UEGM amplitude were most pronounced. The amplitude of the negative peak almost decreased to the baseline, however, it remained negative with an amplitude of -0.96 mV. Along with the amplitude, also the maximum downstroke velocity decreased, which led to a smoother shape of the peak. The described changes could also be identified in the bipolar EGM, which is dominated by the distal EGM for this catheter orientation. Absolute amplitudes were smaller, than in the unipolar case. Quantitatively, this implies a peak amplitude of 5.6 mV vs. 4.5 mV before ablation (0%) and 0.96 mV vs. 0.68 mV at 110% lesion depth.

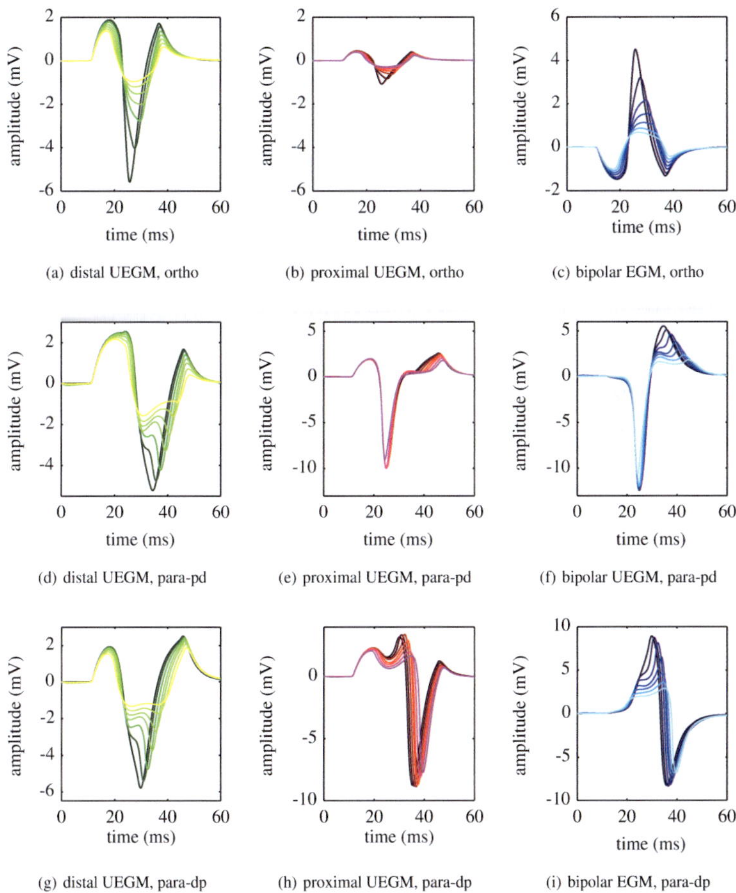

Figure 14.10. Filtered simulated EGMs received from the full lesion model for different catheter orientations; Orthogonal orientation (ortho), parallel catheter orientation with distal to proximal propagation direction (para-dp) and parallel catheter orientation with proximal to distal catheter orientation (para-pd); Color code proximal EGM lesion transmurality: 0% (–), 28% (–), 50% (–),71% (–),92% (–),100% (–),110% (–); Color code distal UEGM: lesion transmuarlity 0% (–), 28% (–), 50% (–), 71% (–), 92% (–), 100% (–), 110% (–); Color code bipolar EGM: lesion transmurality: 0% (–), 28% (–), 50% (–),71% (–),92% (–),100% (–),110% (–)

Parallel Catheter Orientation

Morphology of bipolar EGMs for parallel catheter orientations depends on propagation direction. However, as seen previously, ablation related changes are always more pronounced in the distal ablation electrode signal, which forms the positive peak of the bipolar EGM. With varying orientation, the temporal order of the positive and negative peak differs. With increasing scar volume, the positive peak amplitude decreased, which produced an almost flat appearance of the positive peak. This behavior was also found in the clinical examples in Fig. 14.6 as well as Fig. 14.8.

Along with the bipolar EGM, also the negative peak amplitude of the distal UEGM decreased. Again, a certain flattening of the peak occurred, which is in agreement with the clinical data. In case of a distal to proximal propagation direction, a certain temporal dispersion of the proximal UEGM could be observed.

14.5.3 Quantitative Analysis and Comparison to Clinical Results

Searching for a criterion for lesion assessment, the dependency of characteristic parameters of the level of transmurality is of interest. In Fig. 14.12, relative amplitude changes are plotted against the percentage of lesion transmurality. Changes are displayed for negative peak amplitudes of distal UEGMs and for positive peak amplitudes of bipolar EGMs. Regarding both EGM types, it turned out that changes were similar in distal UEGMs and bipolar EGMs for one catheter orientation. EGM amplitudes decreased in a rather quadratic fashion for orthogonal catheter orientation. In the parallel case, the curve decreased moderately for lesion depths between 0% and 50%. Afterwards, the curve followed a steeper, almost linear decrease.

Relative changes of the positive and negative peak amplitude as well as the maximum derivative between these points are listed in Tab. 14.3. In the clinical data, differences before and after ablation were related to the value before ablation. For the simulated EGMs, lesion levels of 0% and 110% were analyzed. As stated before, negative peak amplitude of distal UEGMs showed most pronounced changes during the clinical and the simulated ablation sequence. In the case of orthogonal catheter orientation, an amplitude decrease

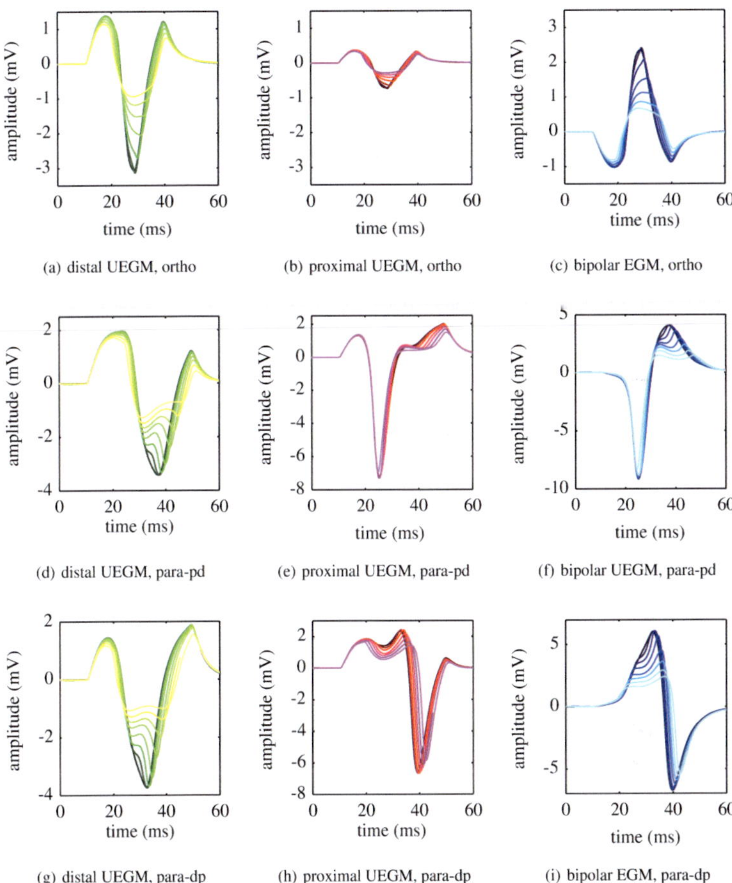

(a) distal UEGM, ortho

(b) proximal UEGM, ortho

(c) bipolar EGM, ortho

(d) distal UEGM, para-pd

(e) proximal UEGM, para-pd

(f) bipolar UEGM, para-pd

(g) distal UEGM, para-dp

(h) proximal UEGM, para-dp

(i) bipolar EGM, para-dp

Figure 14.11. Filtered simulated EGMs received from the full lesion model and additional passive layers on the endo- and epicardial surface; orthogonal catheter orientation (ortho), parallel catheter orientation with distal to proximal propagation direction (para-dp) and parallel catheter orientation with proximal to distal catheter orientation (para-pd); Color code proximal EGM lesion transmurality: 0% (–), 28% (–), 50% (–),71% (–),92% (–),100% (–),110% (–); Color code distal UEGM: lesion transmuarlity 0% (–), 28% (–), 50% (–), 71% (–), 92% (–), 100% (–), 110% (–); Color code bipolar EGM: lesion transmurality: 0% (–), 28% (–), 50% (–),71% (–),92% (–), 100% (–),110% (–)

Table 14.3. Relative changes of characteristic signal parameters for clinical and simulated EGMs during ablation. Values are displayed as mean change $\pm$ standard deviation

	Positive peak amplitude	Negative peak amplitude	maximum dV/dt
clinical distal UEGMs			
orthogonal	-0.17 ± 0.27	-0.94 ± 0.10	-0.80 ± 0.09
parallel, proximal-distal	-0.31 ± 0.21	-1.00 ± 0.12	-0.88 ± 0.10
parallel, distal-proximal	-0.27 ± 0.53	-1.04 ± 0.14	-0.81 ± 0.05
clinical proximal UEGMs			
parallel, proximal-distal	-0.22 ± 0.21	-0.46 ± 0.12	-0.45 ± 0.19
parallel, distal-proximal	-0.15 ± 0.23	-0.09 ± 0.53	0.09 ± 0.41
clinical bipolar EGMs			
orthogonal	-0.95 ± 0.17	-0.20 ± 0.33	-0.79 ± 0.10
parallel, proximal-distal	-0.59 ± 0.56	-0.39 ± 0.03	-0.46 ± 0.13
parallel, distal-proximal	-1.06 ± 0.37	-0.14 ± 0.54	-0.24 ± 0.82
simulated distal UEGMs			
orthogonal	-0.27 ± 0.00	-0.83 ± 0.00	-0.80 ± 0.00
parallel, proximal-distal	-0.13 ± 0.00	-0.65 ± 0.00	0.07 ± 0.00
parallel, distal-proximal	-0.09 ± 0.00	-0.56 ± 0.00	0.02 ± 0.00
orthogonal, only necrotic	-0.24 ± 0.00	-0.74 ± 0.00	-0.70 ± 0.00
simulated proximal UEGMs			
parallel, proximal-distal	-0.07 ± 0.00	-0.10 ± 0.00	0.13 ± 0.00
parallel, distal-proximal	0.03 ± 0.00	-0.01 ± 0.00	-0.02 ± 0.00
simulated bipolar EGMs			
orthogonal	-0.85 ± 0.00	-0.32 ± 0.00	-0.83 ± 0.00
parallel, proximal-distal	-0.51 ± 0.00	-0.10 ± 0.00	-0.09 ± 0.00
parallel, distal-proximal	-0.53 ± 0.00	-0.04 ± 0.00	0.23 ± 0.00
orthogonal, only necrotic	-0.77 ± 0.00	-0.29 ± 0.00	-0.72 ± 0.00
simulated distal UEGMs with connective tissue layers			
orthogonal	-0.22 ± 0.00	-0.75 ± 0.00	-0.56 ± 0.00
parallel, proximal-distal	-0.19 ± 0.00	-0.71 ± 0.00	-0.10 ± 0.00
parallel, distal-proximal	-0.12 ± 0.00	-0.62 ± 0.00	-0.15 ± 0.00
orthogonal, only necrotic	-0.20 ± 0.00	-0.64 ± 0.00	-0.35 ± 0.00
simulated proximal UEGMs with connective tissue layers			
parallel, proximal-distal	-0.11 ± 0.00	-0.14 ± 0.00	0.05 ± 0.00
parallel, distal-proximal	0.05 ± 0.00	-0.06 ± 0.00	0.06 ± 0.00
simulated bipolar EGMs with connective tissue layers			
orthogonal	-0.77 ± 0.00	-0.26 ± 0.00	-0.63 ± 0.00
parallel, proximal-distal	-0.73 ± 0.00	-0.19 ± 0.00	-0.05 ± 0.00
parallel, distal-proximal	-0.62 ± 0.00	-0.12 ± 0.00	0.04 ± 0.00
orthogonal, only necrotic	-0.64 ± 0.00	-0.17 ± 0.00	-0.31 ± 0.00

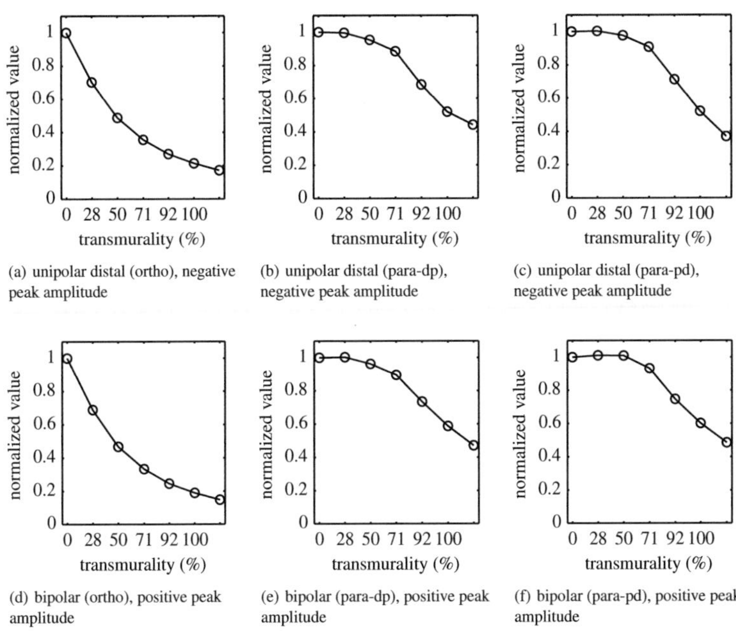

(a) unipolar distal (ortho), negative peak amplitude

(b) unipolar distal (para-dp), negative peak amplitude

(c) unipolar distal (para-pd), negative peak amplitude

(d) bipolar (ortho), positive peak amplitude

(e) bipolar (para-dp), positive peak amplitude

(f) bipolar (para-pd), positive peak amplitude

Figure 14.12. Relative changes of peak amplitudes with regard to lesion transmurality. Values were extracted from EGMs shown in Fig. 14.10. Upper row displays relative changes in negative peak amplitude of distal UEGMs. Bottom row shows relative changes of bipolar positive peak amplitudes. Changes are analyzed for orthogonal catheter orientation (ortho) as well as parallel catheter orientation for distal to proximal propagation (para-dp) and distal to proximal propagation (para-pd).

of 94% for clinical data was measured. This compares to a reduction of 83% for simulated UEGM. Negative peak amplitude values for parallel catheter orientations were shifted in the positive range, which implies a reduction in amplitude of more than 100%. For simulated data a reduction of up to 65% was achieved. Maximum downstroke velocity alterations were reproduced by the simulation in the case of an orthogonal catheter orientation (measured: -80%, simulated: -80%). However, for parallel catheter orientation, simulated upstroke velocity remained almost unchanged, whereas measured EGMs showed a decrease in downstroke velocity. Bipolar EGMs showed similar values for the positive peak amplitude. A peak amplitude reduction of up to 104% was seen in the clinical data, which compares to values of up to 85% in the simulated case. With 53% and 51% the amplitude decrease for parallel catheter orientation was smaller than in the orthogonal case.

In case of simulated EGMs, in which the epicardial surface was covered by a thin layer of passive connective tissue, amplitudes were smaller compared to the case without the epicardial layer. Without lesion (0%), distal UEGM amplitudes were -3.1 mV compared to -5.8 mV for orthogonal catheter orientation. The relative amplitude changes for this set of simulations were less dependent on catheter orientation with relative changes of 62-77% for bipolar EGMs and 62-75% for distal UEGMs compared to 51-85% (bipolar) and 56-83% (unipolar) in the previous case.

Finally, a set of simulations with orthogonal catheter orientation was regarded, in which the lesion was only modeled by a passive necrotic core omitting changes in the border zone. Compared to the full lesion model, relative changes were less pronounced. Amplitude changes in bipolar EGMs were 77% compared to 85% for simulations without connective tissue layers and 64% compared to 77% for the latter case. Similar differences could be identified for unipolar EGMs and the negative downstroke velocity.

14.5.4 Unfiltered EGM Changes During Lesion Formation

This section focuses on signals simulated using the full lesion model with additional connective tissue surface layers. Unfiltered signals for the case without connective tissue layer, which have a similar morphology, can be found in the appendix Fig. A.1.

When comparing signal morphologies of unfiltered simulated EGMs in Fig. 14.13 to their filtered version (Fig. 14.11), severe morphological changes can be found. This agrees with the study of different filter settings, shown in section 10.3.5. However, it is of interest how the changes during ablation are reflected in unfiltered EGMs. Regarding distal UEGMs, which displayed a biphasic morphology before ablation, major changes occurred in the negative peak amplitude and the steepness of the downstroke.

In proximal UEGMs, the representation of lesion effects was orientation dependent. For an orthogonal catheter orientation, the farfield signal changed in a similar way as the distal UEGM. Depending on the location of the proximal electrode relative to the lesion, different parts of the parallel EGMs were altered. If the wavefront reached the lesion area before passing the proximal electrode, the first (positive) part of the EGM showed morphological variation. The opposite was true, if the wavefront reached the lesion area after passing the recording electrode.

Similar behavior was found in bipolar EGMs. When the catheter was oriented orthogonally, both peaks of the biphasic shape gradually decreased. The monophasic morphology variations correlated with the propagation direction. When the wavefront first reached the proximal electrode, the upstroke part following the negative peak became less steep. In the reverse case, the steepness of the upstroke part of the positive peak was reduced.

14.5.5 Translation of Characteristic Signal Points to Excitation Patterns on Tissue Level

Intracardiac EGM analysis tries to extract information from EGM parameters about propagation patterns on the tissue level. Therefore, it is of interest, how EGMs can be analyzed with this respect. Three characteristic points of the wave propagation were chosen as points of interest: The moments when the wavefront

- reaches the lesion border zone,
- passes underneath the distal electrode, and finally
- exits the lesion border zone area.

Corresponding extracellular potential distributions for these events are depicted in Fig 14.14(a-c) for a lesion level of 50%. Time points correspond-

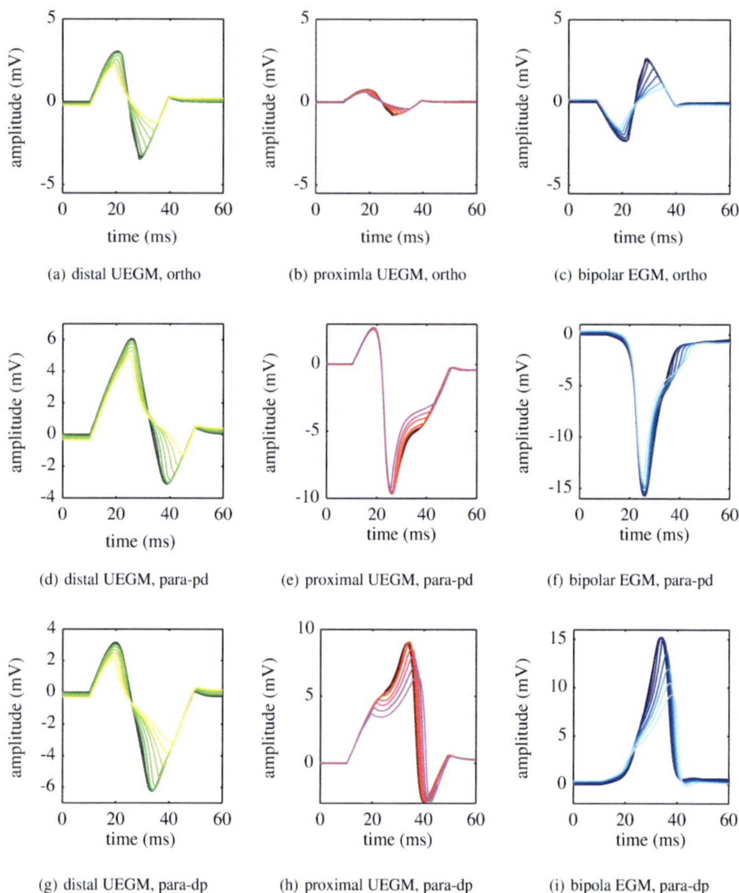

Figure 14.13. Unfiltered simulated EGMs received from the full lesion model and additional passive layers on the endo- and epicardial surface; orthogonal catheter orientation (ortho), parallel catheter orientation with distal to proximal propagation direction (para-dp) and parallel catheter orientation with proximal to distal catheter orientation (para-pd); Color code proximal EGM lesion transmurality: 0% (–), 28% (–), 50% (–),71% (–),92% (–), 100% (–),110% (–); Color code distal UEGM: lesion transmuarlity 0% (–), 28% (–), 50% (–), 71% (–), 92% (–), 100% (–), 110% (–); Color code bipolar EGM: lesion transmurality: 0% (–), 28% (–), 50% (–), 71% (–),92% (–), 100% (–),110% (–)

ing to the events of interest were identified from local activation times. The wavefront arrived at the lesion border zone after a simulated time of 19.5 ms. It passed the distal electrode at 25.9 ms and exited the lesion area at 32.2 ms. Corresponding simulated EGMs are depicted in Fig. 14.14(d-f). The characteristic time points are highlighted by vertical lines. In the unfiltered EGMs, the characteristic points can be linked to the peaks. The passage of the wavefront coincides with the downstroke of UEGMs and the upstroke of the bipolar EGM, respectively. In filtered EGMs, the wavefront passage correlates with negative peaks of the unipolar EGMs.

Parallel catheter orientation is depicted in Fig. 14.15. Here, the fiducial time points were identified for a simulation time of 26.2 ms (arrival at the border zone), 34.0 ms (passage of the distal electrode) and 40.7 ms (exit of the lesion area). Again, peaks and downstroke of the unfiltered distal UEGM can be associated with the identified events. In the remaining leads, the point of arrival was found to coincide with the negative peak of the proximal UEGM, which was also present in the bipolar EGM. However, no meaningful signal representation for the two remaining events of interest could be found.

Motivated by the finding, that the time of occurrence of the negative and positive peak may be correlated with the arrival time of the excitation front at the begin and end of the lesion area, LATs at these points were compared to the corresponding peak times (Fig 14.16). High agreement for all scar depths was found in the case of orthogonal catheter orientation. At the start of the lesion area, a correlation coefficient of 0.927 was reached between the LAT and the peak time curve (Fig. 14.16(a)) . In particular for larger lesions above 50% transmurality, a high correlation was present. The two curves differ by a constant offset value of 1 ms. In a set of four additional simulations, only the scar width was varied, keeping the scar depth at a constant value of 92%. Again, a high agreement was reached (Fig. 14.16(b)). Also in this case, a constant offset was present. However, the curves were perfectly correlated (1.000). At the end of the lesion area, the LAT development was identically

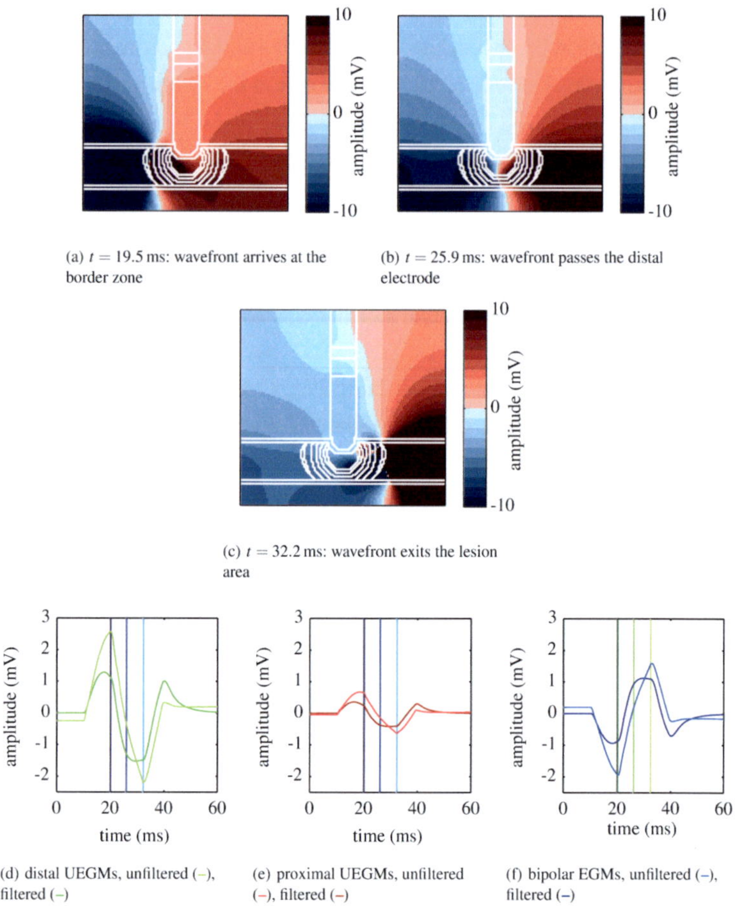

(a) $t = 19.5$ ms: wavefront arrives at the border zone

(b) $t = 25.9$ ms: wavefront passes the distal electrode

(c) $t = 32.2$ ms: wavefront exits the lesion area

(d) distal UEGMs, unfiltered (–), filtered (–)

(e) proximal UEGMs, unfiltered (–), filtered (–)

(f) bipolar EGMs, unfiltered (–), filtered (–)

Figure 14.14. Identification of points of interest in simulated EGMs for an orthogonal catheter orientation. (a-c): Extracellular potential distribution displaying the wavefront location relative to the lesion area. (d-f): Simulated EGMs with vertical lines indicating the time points of the displayed events.

reproduced by the negative peak time (Fig. 14.16(d,e)). This was true for the correlation coefficient (0.999) as well as absolute values. When the catheter was oriented parallel to the tissue surface, the positive peak times remained constant for smaller lesions. Only for lesion depths above 92% peak times

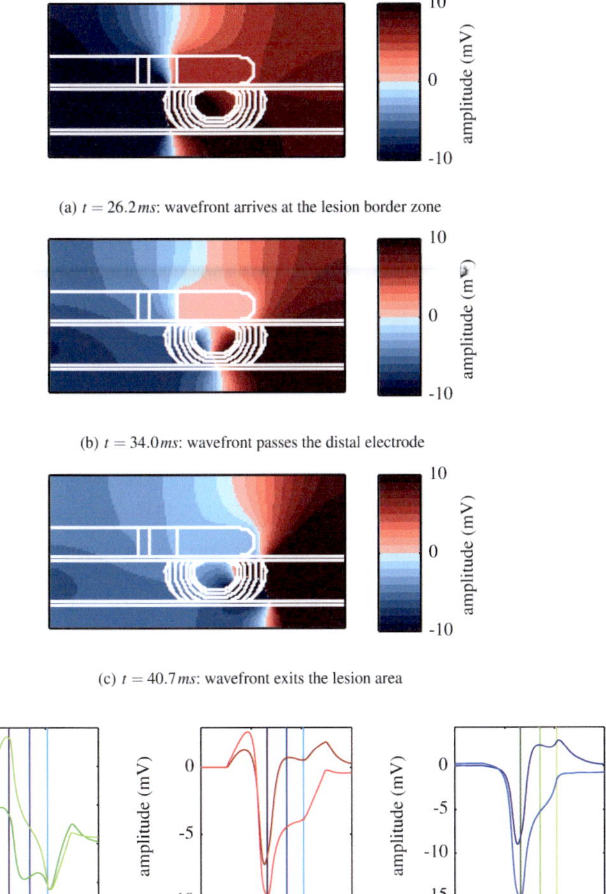

(a) $t = 26.2\,ms$: wavefront arrives at the lesion border zone

(b) $t = 34.0\,ms$: wavefront passes the distal electrode

(c) $t = 40.7\,ms$: wavefront exits the lesion area

(d) Distal UEGMs, unfiltered (–), filtered (–)

(e) Proximal UEGMs, unfiltered (–), filtered (–)

(f) Bipolar EGMs, unfiltered (–), filtered (–)

Figure 14.15. Identification of points of interest in simulated EMGs for a parallel catheter orientation. (a-c): Extracellular potential distribution displaying the wavefront location relative to the lesion area. (d-f): Simulated EGMs with vertical lines indicating the time points of the displayed events.

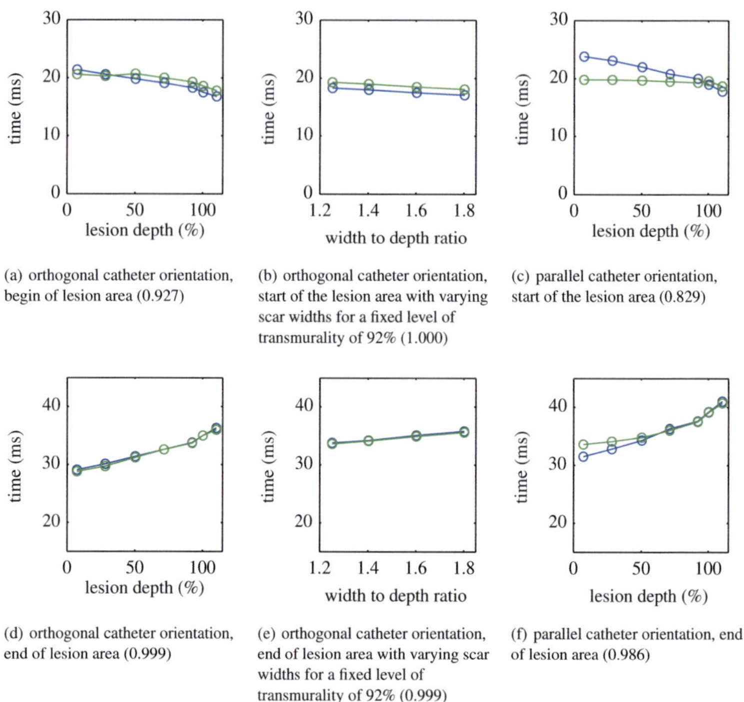

(a) orthogonal catheter orientation, begin of lesion area (0.927)

(b) orthogonal catheter orientation, start of the lesion area with varying scar widths for a fixed level of transmurality of 92% (1.000)

(c) parallel catheter orientation, start of the lesion area (0.829)

(d) orthogonal catheter orientation, end of lesion area (0.999)

(e) orthogonal catheter orientation, end of lesion area with varying scar widths for a fixed level of transmurality of 92% (0.999)

(f) parallel catheter orientation, end of lesion area (0.986)

Figure 14.16. Comparison of the time points of negative and positive peaks in the distal UEGM (–) with the Local activation times at the start and end of the lesion area (–). Values in brackets show the (correlation coefficient) between the two curves.

reproduced the LATs. Regarding the end of the lesion area, a good fit for lesions larger or equal 50% was obtained.

14.6 Discussion

In this chapter, EGM changes during radiofrequency ablation were studied. A simulation model for RFA lesions in the acute stage was developed. This model accounts for the ellipsoidal shape of RFA lesions inside the cardiac wall. Furthermore, it includes acute electrophysiologic characteristics of RFA lesions, which are found in a border zone surrounding the necrotic lesion core. Simulated EGMs reproduced characteristic changes, which were identified in clinically measured human intracardiac EGMs. The presented simulation approach was parametrized for the bidomain model of cardiac excitation. Therefore, the influence of a large ablation electrode on cardiac electrophysiology is taken into account. Furthermore, it allows for extracellular stimulation, which is a frequently used technique in clinical procedures to evaluate lesion completeness.

Until today, several simulation studies have addressed radio frequency ablation lesions in the context of atrial fibrillation [169–173]. Their objective was the influence of lesion patterns on cardiac spread of excitation. However, these studies did not present intracardiac EGMs in scar regions. Furthermore, scars in these studies were only modeled as passive necrotic areas, which is a valid representation for the chronic state of the lesion, days and weeks after the ablation therapy. Simulated intracardiac EGMs, which were not studied in the context of RFA, were presented in [117, 143–145, 160, 174]. In contrast to these studies, in this work a geometrically detailed model of a clinical ablation catheter was used, which is represented by highly conductive electrodes.

Experimental Investigation of CV During Elevated Temperatures

Conduction velocities were calculated based on the distance between barycenters of LAT areas. Also gradient based methods have been investigated [175, 176] for CV estimation. However, the low temporal resolution and noisy optical mapping data led to significant errors in estimated CVs using these methods. Optical measurements of CV during a temperature profile between $36.7\,^{\circ}\mathrm{C}$ and $43.8\,^{\circ}\mathrm{C}$ were investigated. CV was increased up to 119% of baseline CV at $42\,^{\circ}\mathrm{C}$ showing a decrease when reaching $43.8\,^{\circ}\mathrm{C}$. These findings confirm the temperature profile recorded by Simmers et. al., who also found a maximum CV around $42\,^{\circ}\mathrm{C}$ with a value of 114%. An increase of CV in an RFA lesion borderzone was also reported in [52]. This data further supports the choice of elevated temperature in the wider lesion borderzone of the presented simulation model. When returning to baseline temperature, CV remained elevated by 4%, this finding corresponds to the study in [52], where RFA related changes were evaluated between 1 min and 30 min after RF application. Based on this result it might be of interest to study temporal development of tissue recovery after RF application. The time necessary to return to baseline values may be an important parameter for lesion assessment in clinical practice.

Clinical Data

Eleven datasets measured in five patients showing intracardiac EGMs during ablation were analyzed for characteristic changes due to the lesion formation process. The main finding was a mean amplitude reduction of 98% of the negative peak in distal UEGM. This was linked to a reduced downstroke velocity. As a consequence, the part of bipolar EGMs, which is dominated by the distal UEGM, was reduced in the same fashion. In proximal EGMs, only minor changes occurred. Due to the variable morphology of bipolar EGMs, distal UEGMs seem to be more suitable for lesion assessment. The found changes are in agreement with the results from a porcine model [6] and a multi patient study of human EGMs [168].

A novel and important finding of the study in this work is the temporal development of the reported amplitude changes during and after the ablation process. Major EGM changes were limited to the time frame of 5-12 s after ablation onset for the first ablation in one location. This is in agreement with the temporal characteristics of lesion development reported from ultrasound measurements [177] and histological data [135]. In 10 of the 11 ablation sequences, signal morphology was stable after RFA application for the next 20 s. However, one example was presented, in which the signal amplitude partially recovered after the first ablation sequence. After another RF application, the same reversible behavior was detected. The presented example was the only case, in which data was unambiguous. More examples were found, in which a similar behavior might be present. However, for these cases, other influencing parameters like unstable electrode position could not be ruled out. A possible interpretation of this phenomenon may be reversible effects, which were also reported in experimental data [49]. The implication for clinical application would be, that EGMs do not have to be studied only during RF application, but also a certain time preceding the ablation sequence to exclude lesion recovery.

Acute RFA Lesion Model

The applied parametrization of the model of acute RFA lesions was based on the sparse literature data covering ablation related changes in cardiac tissue. The ellipsoidal lesion shape was derived from simulation and experimental data presented in [6, 45–47, 136, 178, 179]. The electrophysiological changes, which were introduced in section 3, were macroscopically reproduced by the modified ten Tusscher cell model adapted to ischemia phase 1a [105]. However, underlying mechanisms leading to these changes are not completely understood. Among the possible mechanisms, cytosolic calcium levels [56] and ischemic effects due to reduced perfusion [40] are discussed. The adaptation of CV parameters by modified intracellular conductivities can be justified by changes in gap junction characteristics during hyperthermia [55]. Thereby, a temperature dependent CV gradient resembling the data of [50] was created.

Filtered Simulated EGMs

A key parameter to fit simulated data to clinical electrograms was the exact knowledge of the filter function used by the clinical recording system. The simulation environment was used in various degrees of complexity to study EGM changes during lesion development. Three components of the simulation environment were distinguished:

- The necrotic lesion core
- A lesion border zone with changed AP and CV characteristics
- Connective tissue layers on the endo- and epicardial surface of the myocardial patch

By this differentiation, the sensitivity of the individual parameters could be assessed. In a stepwise application of different lesion parameters, it was found that using a necrotic core with an ellipsoidal lesion shape, a characteristic reduction in peak amplitudes could be obtained. However, this would be far off the changes seen in clinical data. By including the lesion border zone, the relative amplitude change was more pronounced and the downstroke velocity decreased. It was found, that the CV characteristics had a larger impact on EGM shape, than AP changes. When connective tissue layers were applied, which are supposed to account for the deviating properties of the endo- and epicardium, an overall reduction in EGM amplitudes could be observed. When accounting for this feature, the relative amplitude change for orthogonal catheter orientation decreased, whereas this parameter increased for parallel catheter orientation. Similar behavior was found for the downstroke velocity. Overall, it can be stated that all parameters contribute to EGM shape and amplitude and can therefore be optimized to fit certain data. The focus of this study was to rely on the given experimental data without accounting for possible variations.

The simulated EGMs achieved good morphological agreement with the clinical data for the part, at which the wavefront passed the catheter electrodes. This so called near field part is of major interest, as it is also subject to ablation related changes. The onset of simulated EGMs commenced with a relatively steep rise, which was produced by the planar stimulus, that created a planar wavefront relatively close to the catheter electrodes. Although this poses a difference to clinical data, it should not influence the outcome of the study, which focussed on changes in the nearfield part. This should be true for unfiltered signals as well as filtered versions, as the impulse response of the first order Butterworth filters is short. Resulting amplitude changes in simulated data were smaller than seen in clinical data. Various reasons may contribute to this difference. On the one hand it was previously mentioned, that all lesion components influence EGM amplitude and could be adjusted to modify this parameter. Furthermore, the model does not account for the effects present at the electrode tissue interface. At this position electrochemical effects due to ablation currents may lead to low frequency DC effects [90]. Finally, it was found that not only lesion depth, but also lesion width increase leads to reduced EGM amplitudes.

Unfiltered EGMs and Correspondence to Myocardial Excitation

Regarding unfiltered EGMs the best representation of ablation related changes was also found in the distal UEGM. Additionally to the negative peak amplitude, also the downstroke velocity was strongly reduced. In addition to peak amplitudes, a further criterion for lesion assessment was found in the point in time, when the positive and negative peaks were formed. These points showed a high correlation with the time, when the wavefront reaches and leaves the lesion area. Therefore, the peak times could be used as a marker to estimate lesion width.

An important question of this simulation study is the implication of the results for the assessment of lesion transmurality based on EGM based parameters. No sharp characteristic change could be identified, when reaching complete lesion transmurality. Although simulated EGMs did not present complete loss of the negative peak amplitude, there is no indication, that the gradual peak amplitude reduction should be different in real ablation situations. In simulations, the peak amplitude slowly decreased, before the lesion reached the transmural state and continued decreasing, when the lesion grew further in vertical direction. Based on these findings, it can be stated, that negative peak amplitude of unipolar electrograms can be used as a criterion for lesion control. However, the temporal development of this marker may be regarded with higher attention. If the lesion stops growing, this may be reflected in the change of this parameter. Furthermore, a recovery of the peak amplitude may indicate reversible changes.

15

Simulation of Optimized Catheter Designs

In the previous chapter a simulation environment for acute ablation lesions was introduced. The current chapter presents simulation results of possible future ablation catheter designs, which are investigated regarding their performance in EGM based lesion assessment. One catheter has three microelectrodes (*Micro-Catheter*) included in the ablation electrode for a higher local resolution. The second catheter is formed by a centered ablation electrode surrounded by four measurement electrodes (*Cross-Catheter*). Both catheters were described in detail in section 9.1.1. The current chapter is focussed on bipolar EGM combinations, which should bear additional information due to the geometric settings of the catheter electrodes.

15.1 Applied Methods

Microelectrode Ablation Catheter (Micro-Catheter)

In order to model the three microelectrodes inside of the ablation tip electrode a spatial resolution of 0.06 mm is needed. Therefore, a smaller simulation setup had to be used to keep simulation in an affordable timespan. Nevertheless, still six days were needed for one EGM simulation on eleven parallel cores. The dimensions of the simulated piece of cardiac tissue were set to 12 mm×12 mm×3 mm. Simulated stages were 0%, 40%, 80% 100% and 110% transmural extent of the lesion with a width to depth ratio of 1.25, accounting for the orthogonal catheter orientation.

Figure 15.1. Simulation setup showing the *Micro-Catheter* on a planar patch of cardiac tissue (green). Underneath the catheter a lesion area is outlined in blue. Planar stimulation was applied at the left side of the shown patch. Microcatheter electrodes, which form a triangle of equal side lengths are numerated according to their LAT (not shown). The axis between electrodes 1 and 3 is aligned with the x-axis of the simulation setup and therefore with the propagation direction.

Catheter with a Centered Ablation Electrode (Cross-Catheter)

Using the described approach to model acute ablation lesions (Sec. 9.1.4) five stages of transmurality were simulated (0%, 20%, 40%, 80%, 120%) on a planar myocardial patch (32 mm×24 mm ×5 mm) with a spatial resolution of 0.2 mm. Lesion width to depth ratio was set to 1.6. No passive endo- and epicardial surface layers were included to minimize the complexity of the model. The *Cross-Catheter* (Sec 9.1.1) was positioned parallel to the tissue surface, in a way that all measurement electrodes were in contact with the tissue. Due to the higher electrode diameter of the ablation electrode it was indented into the tissue surface by 1 mm, whereas the four measurement electrodes were touching the tissue surface without deformation (Fig. 15.2).

15.2 Simulation Results for the *Micro-Catheter*

The *Micro-Catheter* provides various bipolar and unipolar signal combinations. First bipolar EGMs between the microelectrodes are introduced (Fig 15.3), subsequently unipolar EGMs (Fig 15.4) and bipolar EGMs between microelectrodes and the ablation electrode (Fig 15.5) are evaluated.

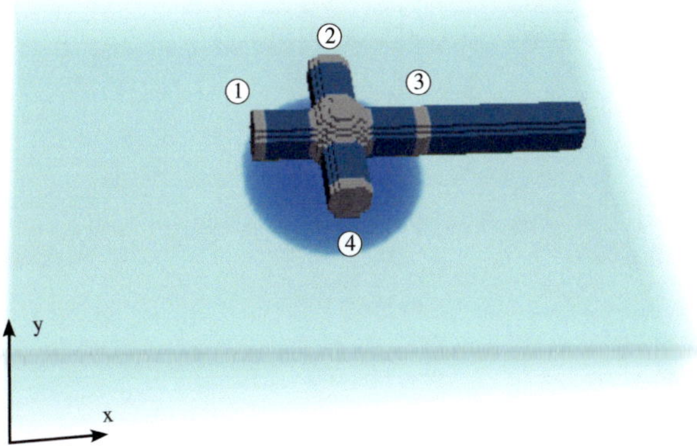

Figure 15.2. Simulation setup showing the cross ablation catheter on a planar patch of cardiac tissue (green). Underneath the catheter a lesion area is outlined in blue. Catheter electrodes are numbered clockwise starting with the electrode, which is reached first by the planar wavefront. Planar stimulation was applied at the left side of the patch.

Furthermore, quantitative results for the negative peak amplitude and the maximum negative deflection are displayed in Fig. 15.6. Due to the similar development only one exemplary curve is given for each group of signals. Although, being completely covered by the necrotic core, bipolar microelectrode signals show a gradual negative peak amplitude decrease over all lesion stages. The original triphasic morphology became biphasic with a more pronounced S-wave. Quantitatively, the negative peak amplitude of EGMs recorded between microelectrodes 1 and 3 decreased linearly to 80% of the baseline value for transmural lesions. Both negative and positive peak amplitudes of unipolar EGMs (Fig 15.4) were significantly smaller when reaching transmurality. In the EGMs of microelectrode 3 a characteristic change was observed for the two transmural lesion stages. These two electrograms showed a completely positive monophasic morphology compared to a biphasic one for non-transmural cases. Similar EGM characteristics were observed in bipolar EGMs between the ablation electrode and one of the three microelectrodes (Fig 15.5). Again, in the signal of microelectrode 3 a monophasic positive EGM was obtained. Quantitative negative amplitude

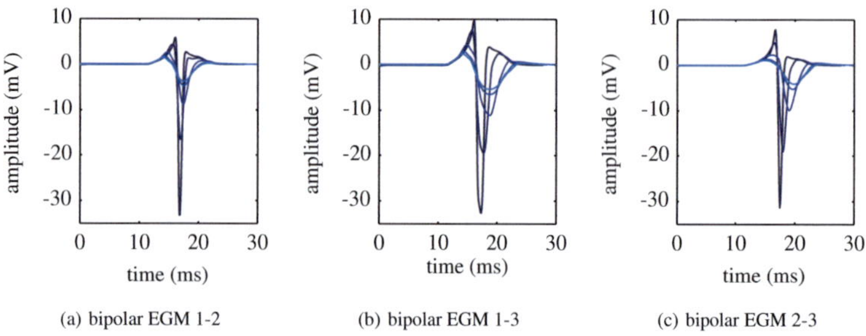

(a) bipolar EGM 1-2   (b) bipolar EGM 1-3   (c) bipolar EGM 2-3

Figure 15.3. Bipolar electrograms between the three microelectrodes of the *Micro-Catheter*. Lesion transmurality: 0% (–), 40% (–), 80% (–),100% (–),110% (–)

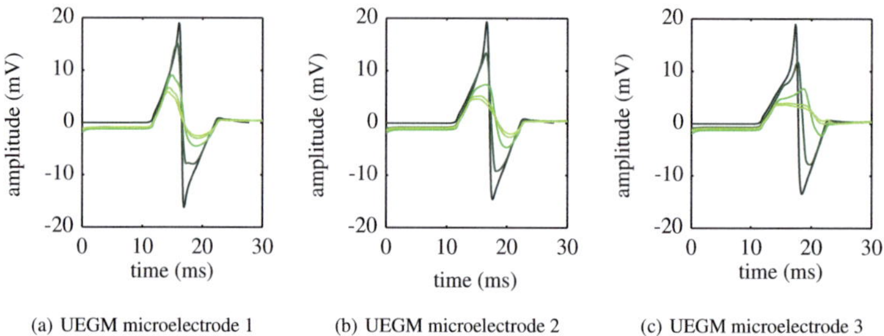

(a) UEGM microelectrode 1   (b) UEGM microelectrode 2   (c) UEGM microelectrode 3

Figure 15.4. Simulated unipolar electrograms of the three microelectrodes of the *Micro-Catheter*. Lesion transmurality: 0% (–), 40% (–), 80% (–),100% (–),110% (–)

curves showed a linear shape up to 100% transmural lesion extent with a loss of the negative peak for transmural lesions (Fig. 15.6). Quantitiative analysis of the maximum first derivative of the simulated EGMs produced a non-linear declining curve. In EGMs measured on larger lesions, this parameter attained only 3-5% of the original value.

15.3 Simulation Results Using the *Cross-Catheter*

Due to lesion formation a reduction in maximum positive or negative peak amplitude was observed. Additionally electrogram width increased (Fig. 15.7). In the first simulation steps several EGM combinations (0-1, 0-3, 2-3, 1-3, 1-2) displayed a pronounced first peak followed by a small second peak

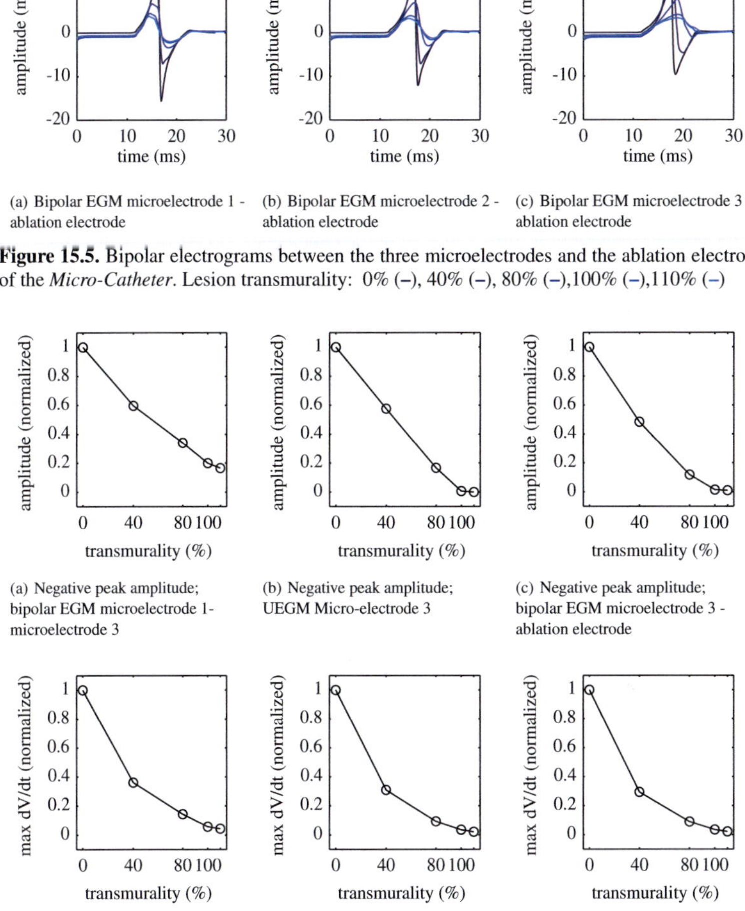

(a) Bipolar EGM microelectrode 1 - ablation electrode

(b) Bipolar EGM microelectrode 2 - ablation electrode

(c) Bipolar EGM microelectrode 3 - ablation electrode

Figure 15.5. Bipolar electrograms between the three microelectrodes and the ablation electrode of the *Micro-Catheter*. Lesion transmurality: 0% (–), 40% (–), 80% (–),100% (–),110% (–)

(a) Negative peak amplitude; bipolar EGM microelectrode 1- microelectrode 3

(b) Negative peak amplitude; UEGM Micro-electrode 3

(c) Negative peak amplitude; bipolar EGM microelectrode 3 - ablation electrode

(d) Maximum deflection; bipolar EGM microelectrode 1- microelectrode 3

(e) Maximum deflection; UEGM Micro-electrode 3

(f) Maximum deflection; bipolar EGM microelectrode 3 - ablation electrode

Figure 15.6. Development of the negative peak amplitude and maximum deflection in exemplary signals of the *Micro-Catheter*

of inverted polarity. This slightly biphasic morphology changed to a purely monophasic one with increased lesions size. EGM 0-1 switched from biphasic to monophasic between 40% and 80% transmurality. In other leads this morphological change first occurred upon lesion transmurality. A remarkable morphological development was observed for EGM 0-2, which was calculated between the ablation electrode and the electrode next to it, which was reached by the wavefront at the same time. In the case without lesion it displayed a small biphasic shape with positive to negative deflection. During the ablation process its amplitude was reversed with stepwise changing amplitudes.

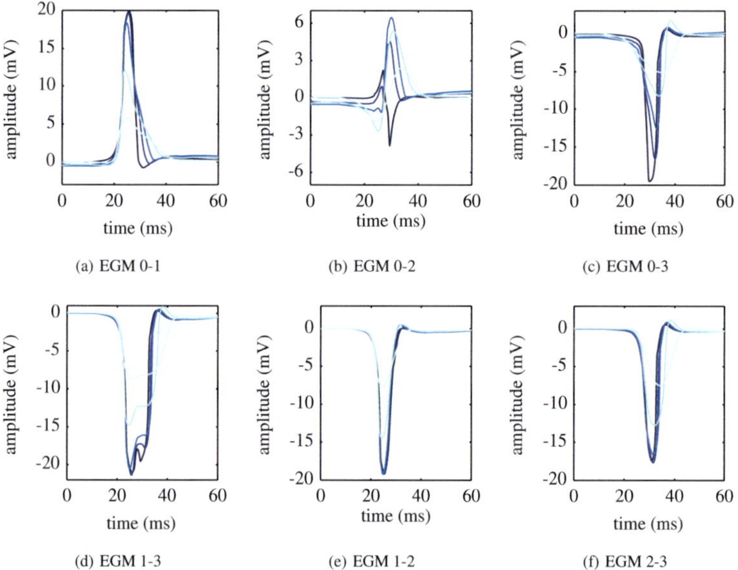

Figure 15.7. Simulated bipolar signals of the *Cross-Catheter*. Nomenclature for EGM X-Y implies a bipolar EGM, where the signal was obtained by the extracellular potential on electrode X minus the potential on electrode Y. Lesion transmurality: 0% (–), 20% (–), 40% (–), 80% ( ), 120% ( ).

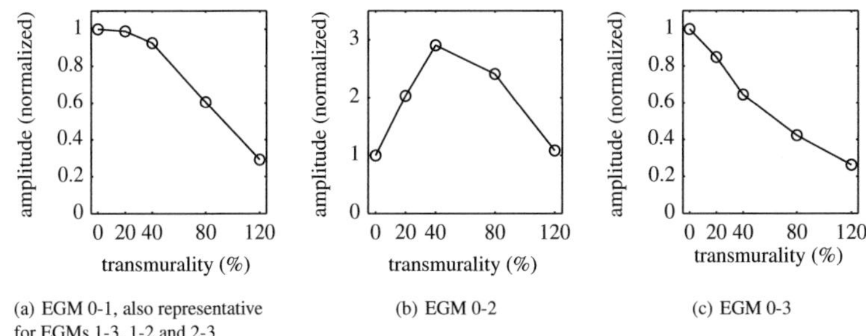

(a) EGM 0-1, also representative
for EGMs 1-3, 1-2 and 2-3

(b) EGM 0-2

(c) EGM 0-3

Figure 15.8. Normalized EGM amplitude values with respect to lesion transmurality for different bipolar combinations introduced in Fig. 15.7

EGM width and amplitude are candidates for robust markers in clinical diagnosis. A quantitative analysis is shown in Fig. 15.8 and Fig. 15.9. Maximum peak amplitude of EGMs 0-1, 1-3, 1-2, and 2-3 at the end of the ablation sequence was 30-40% of the amplitude without lesion. EGM 0-2 signal amplitudes first increased after ablation onset and then decreased to the baseline level. EGM 0-3 showed an approximately linear course with increasing lesion extension. In EGM combinations between the centered ablation electrode and one peripheral electrode (EGM 0-1, 0-2, 0-3) EGM width increased with increasing lesion size between 300-500% of the baseline value. EGM width increase was smaller in EGM combinations between peripheral electrodes (EGM 1-3, EGM 1-2, EGM 2-3), which reached values of 160-250% . Clinical visual diagnosis may be supported by a combination of multiple electrodes. In Fig. 15.10(a) an exemplary signal composed of the sum of EGM 0-1 and EGM 0-3 is shown. In this combined signal, changes add up, which can also be seen in the quantitative analysis (Fig. 15.10(b), (c)). In the presented combination peak-to-peak amplitude decreased linearly with respect to lesion size. The electrogram width parameter was not altered by the summation of signals.

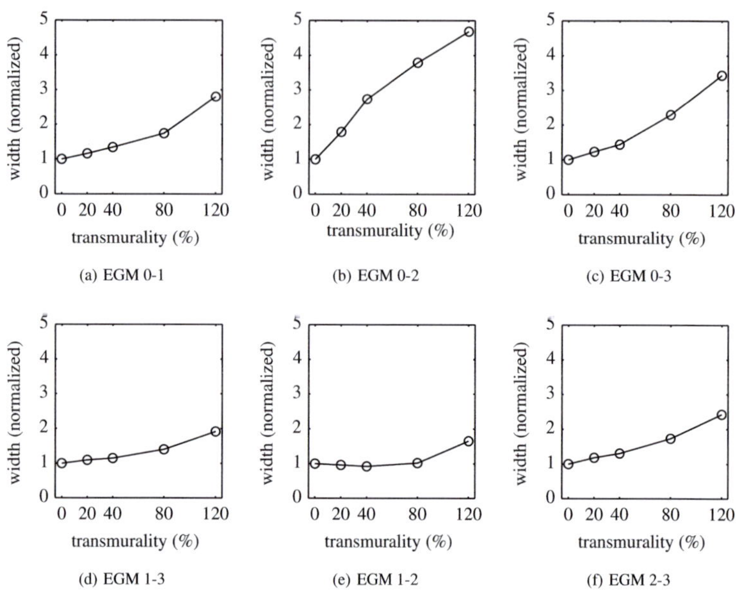

(a) EGM 0-1   (b) EGM 0-2   (c) EGM 0-3

(d) EGM 1-3   (e) EGM 1-2   (f) EGM 2-3

Figure 15.9. Normalized EGM width values with respect to lesion transmurality for different bipolar combinations introduced in Fig. 15.7

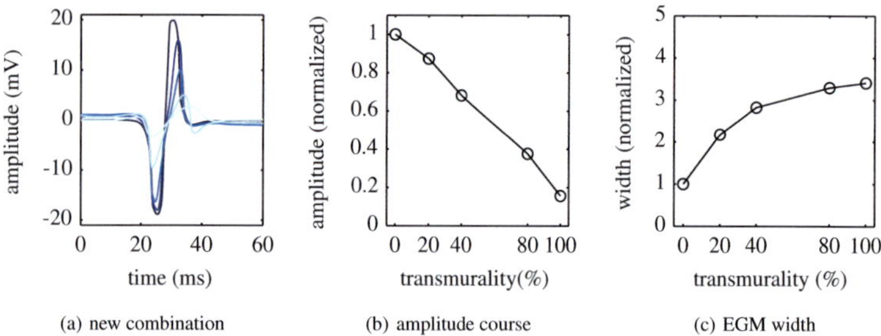

(a) new combination   (b) amplitude course   (c) EGM width

Figure 15.10. Combined signal formed by the sum of EGM 0-1 and EGM 0-3. Lesion transmurality: 0% (–), 20% (–), 40% (–), 80% (–), 120% ( ). The amplitude decrease is approximately linear over the whole lesion development.

15.4 Discussion

In this study two potential future catheter designs were investigated. On the one hand three micro-electrodes were included in the catheter tip. On the other hand, four peripheral electrodes were positioned around a centered electrode. Both catheters showed possibly useful aspects.

Signals of the microelectrodes of the *Micro-Catheter* in different combinations showed a stronger amplitude reduction during lesion formation than the conventional catheter presented in the previous chapter. Especially, unipolar EGMs and bipolar EGMs between one of the microelectrodes and the ablation electrode were sensitive to this parameter. In microelectrode 3, which was reached last by the excitation front, morphology of the signal switched to monophasic, when transmurality was reached. The simulation results are supported by the findings shown in [180], where an ablation catheter with four microelectrodes at the lateral side of the ablation electrode was investigated. This study also reports more pronounced amplitude changes on the microelectrode signals compared to the signals recorded from the tip-ring combination.

Amplitude changes in signals recorded by the cross catheter were similar to those of the conventional catheter in parallel orientation (-60 to -70%) for electrode combinations in line with propagation direction. However, analysis of bipolar signals between electrodes at the side of the catheter with respect to the ablation electrode revealed strong amplitude and morphological changes during lesion development. Multiple electrodes surrounding the catheter electrode may provide information of lesion development in a specific lateral direction. Finally, a combination of three electrode signals displayed a linear amplitude change with increasing lesion size. Two bipolar electrograms of electrodes in front and behind the ablation electrode with respect to propagation direction were added up. In particular, for visual online assessment such an electrode combination may be advantageous.

16

Summary and Conclusions

16.1 Summary

The focus of this thesis was an improved understanding of intracardiac electrogram morphology. A multidisciplinary approach combining in-vitro experiments, clinical data analysis and simulations of cardiac electrophysiology was pursued.

Experimental Setup for Electrical and Optical Measurements of Cardiac Electrophysiology

A new experimental setup was developed, which allows for simultaneous optical and electrical acquisition of cardiac electrophysiological data. Combining both modalities, it is possible to acquire electrical extracellular recordings under defined conditions. For electrical measurements a new sensor was developed, which resembles electrode spacing and lengths of clinical mapping catheters. It was used to acquire electrograms in unipolar and bipolar configurations. Precise positioning of the sensor was enabled using a computer controlled micro-manipulator. Furthermore, an optimized tissue bath was developed, in which vital preparations can be accessed from above for electrical measurements and optically observed at the same time under predefined temperature conditions. Several limitations of in-vivo measurements could be overcome. On the one hand, position and orientation of the measurement electrode relative to the tissue can be precisely determined. On the other hand, underlying patterns of excitation propagation can be studied by optical recordings. In this work, cases of electrical and optical recordings on papillary muscles and right atrial preparations were presented, in which also the relationship of electrode-tissue distance was investigated. Electrogram characteristics could be interpreted with the knowledge of underlying excitation patterns. Furthermore, conduction velocity development under hyperthermic conditions was studied.

Analysis of Clinical Intracardiac Electrograms

Five annotated datasets of clinically measured intracardiac electrograms were analyzed. A method for template generation was implemented, which enabled a detailed study of electrogram morphology. Thereby, morphological changes due to changing catheter orientation could be identified. Furthermore, detailed datasets on morphological changes during radio frequency ablation could be investigated.

High Resolution Simulation Framework for Intracardiac Electrogram Studies

A new simulation framework incorporating geometrically detailed models of clinical measurement catheters was established. This was achieved, by modeling catheters including their highly conductive electrodes and the separating isolations. Using the detailed bidomain model to describe excitation propagation, interactions between the highly conductive electrodes and electrophysiological parameters could be investigated. The main application of this simulation setup is the detailed simulation of the nearfield part of the electrogram, which is formed, when an excitation wavefront passes underneath the measuring electrodes.

Applications

Experimental and clinical data allowed for a verification of the simulation environment under physiological conditions. In detailed simulations the corresponding measurement situations were reproduced. High agreement was obtained in clinical EGM morphology for different catheter orientations. The amplitude development measured experimentaly in relation to electrode-tissue distance was closely reproduced in the simulation environment. The most important finding of this study was a sigmoidal shaped curve describing the relationship between peak-to-peak amplitude and electrode tissue distance. Based on simulated data the inflection point of the curve was associated with first contact between electrode and tissue. Experimental data showed a further increase in amplitude, when the electrode was indented into the tissue, which formed the second part of the sigmoidal function.

The validated simulation approach enabled detailed studies of influencing parameters on electrogram morphology under physiological conditions. It was found, that catheter orientation strongly changes morphology and amplitude of bipolar electrograms. Furthermore, EGM changes in relation to cardiac wall thickness and conduction velocity were assessed. This data provides comprehensive insights for electrophysiologists and can therefore lead to improved electrogram interpretation in clinical procedures.

Consequently, the spectrum of applications was extended to pathological substrates. First, the influence of three-dimensional fibrotic substrates on electrogram morphology was evaluated. The focus of this study was the representation of different fibrotic patterns (patchy, diffuse) in the signals measured by clinical mapping catheters, in order to contribute to the understanding of complex fractionated atrial electrograms (CFAE). Independent of the type of fibrosis, conduction velocity decreased linearly with increasing fibrotic volume fraction. Electrogram fractionation only occurred on patterns of patchy fibrosis, which was formed by longer, strand like fibrotic elements. Furthermore, electrogram fractionation was dependent on propagation direction relative to the orientation of fibrotic elements. Electrograms were only fractionated in the case of transversal propagation. Also, an increasing degree of fractionation was observed for larger electrode spacings. Simulated CFAEs were compared to a database of clinical signals. Good agreement was found regarding the amplitudes, the number of deflections and the time between peaks within one electrogram.

In a second study, electrogram morphology was investigated during the creation of radiofrequency ablation lesions. Based on literature data, a model representing ablation lesions in their acute state was parametrized. Surrounding the necrotic lesion core, a borderzone with altered conduction velocity and action potential morphology was included. In an own experimental study, the reported finding, that conduction velocity increases up to a temperature of $42\,°C$ followed by a decrease for higher temperatures could be confirmed. A detailed comparison to clinical electrograms was performed, which showed the ability of the simulation model to reproduce relevant morphological electrogram changes. It was found, that EGM peak amplitudes show a high correlation with lesion size. Furthermore, close agreement between peak times and lesion width was revealed. Simulated electrograms did not show a characteristic sharp change upon reaching lesion transmurality. However, the temporal development of the peak amplitude may still be a good marker for lesion size estimation. Additionally a major outcome of the clinical data analysis was, that monitoring the EGM amplitude after the ablation may help to determine reversible changes in the ablated area and to limit the necessary ablation time. The established lesion model opens the possibility to develop criteria for electrogram based lesion assessment for complex lesion morphologies.

Two optimized catheter geometries were studied regarding their ability to evaluate ablation lesion formation. A catheter with three microelectrodes, included in the ablation electrode tip, provided more pronounced amplitude changes with increasing lesion depth. In two electrode configurations even a characteristic change of electrogram morphology from biphasic to monophasic was determined upon reaching lesion transmurality. A catheter with a central ablation electrode, which is surrounded by four peripheral measurement electrodes showed improved characterization of the lesion width in all directions. Furthermore, multi-electrode combinations could be used to emphasize characteristic morphological changes.

16.2 Perspective

The main outcomes of this work are an experimental setup to study electrogram formation under defined conditions as well as a validated simulation framework to evaluate substrate related changes in intracardiac electrogram morphology. Combining both modalities can support the transition from annotation based clinical diagnosis to detailed analyses of electrogram morphology. A better understanding of the representation of excitation patterns in electrogram morphology may lead to improved clinical diagnosis and therapies. A second important application of the developed methods is the design of new catheters. Like in many other fields of industrial development, the simulation based optimization of new catheters can lead to improved, cost effective solutions.

A

Appendix

A.1 Polynomial Fits Resembling EGM Amplitude and Width Development

EGM Peak-to-Peak Amplitude in Relation to Tissue Thickness

The shown curves in Fig. 10.10 were extrapolated by a polynomial fit of the type:

$$A_{pp}(x_1, x_2, x_3, D) = \frac{x_1 \cdot D + x_2}{D + x_3} \tag{A.1}$$

Corresponding parameters are given in Tab. A.1

Table A.1. Polynomial fit parameters for the dependency of signal peak-to-peak amplitude and myocardial wall thickness

	x_1	x_2	x_3
orthogonal orientation			
distal	10.29	-10.34	14.73
proximal	7.704	-9.128	21.08
bipolar	4.232	-5.456	6.920
parallel orientation			
distal	10.85	-10.24	11.97
proximal	10.470	-7.888	8.373
bipolar	11.28	-12.34	6.76

EGM Peak-to-Peak Amplitude in Relation to Conduction Velocity

Simulated values for A_{pp} in Fig. 10.11 were fitted by a polynomial fit of the type:

$$A_{pp}(x_1, x_2, x_3, x_4, CV) = \frac{x_1 \cdot CV + x_2}{CV^2 + x_3 \cdot CV + x_4} \tag{A.2}$$

Corresponding coefficients are given in Tab. A.2

Table A.2. Polynomial fit parameters for the dependency of signal peak-to-peak amplitude of conduction velocity

	x_1	x_2	x_3	x_4
orthogonal orientation				
distal	2.576	-0.252	-2.470	1.853
proximal	1.358	-0.128	-2.432	1.752
bipolar	1.568	-0.153	-2.428	1.882
parallel orientation				
distal	3.191	-0.315	-2.459	1.876
proximal	4.120	-0.397	-2.484	1.980
bipolar	4.687	-0.475	-2.459	1.960

Fit curves for the dependency between the electrogram width (W) and CV (Fig. 10.11) were produced by the following polynomial expression.

$$W(x_1, x_2, x_3, CV) = \frac{x_1 \cdot CV + x_2}{CV + x_3} \tag{A.3}$$

Corresponding coefficients are given in Tab. A.3

Table A.3. Polynomial fit parameters for the dependency of signal width of conduction velocity

	x_1	x_2	x_3
orthogonal orientation			
distal	0.9522	36.47	-0.0337
proximal	-6.544	53.01	0.083
bipolar	8.464	17.000	-0.054
parallel orientation			
distal	1.060	35.780	-0.0341
proximal	0.600	35.370	-0.035
bipolar	5.534	3.006	-0.121

A.2 Unfiltered Simulated Signals During Lesion Development

The following figure displays unfiltered simulated signals for the full model of acute lesions.

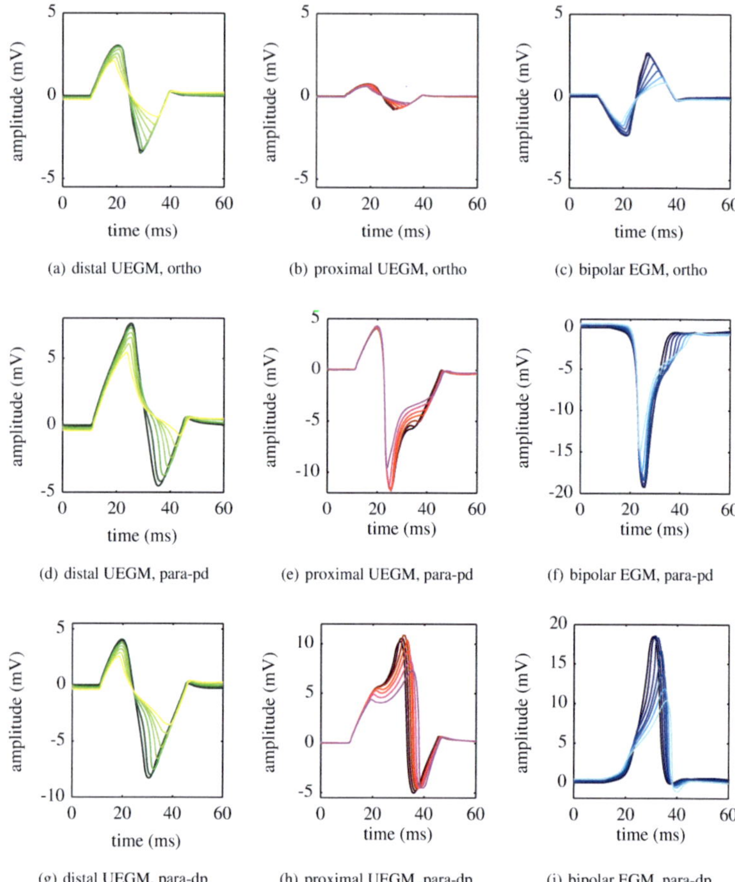

(a) distal UEGM, ortho

(b) proximal UEGM, ortho

(c) bipolar EGM, ortho

(d) distal UEGM, para-pd

(e) proximal UEGM, para-pd

(f) bipolar EGM, para-pd

(g) distal UEGM, para-dp

(h) proximal UEGM, para-dp

(i) bipolar EGM, para-dp

Figure A.1. Unfiltered simulated EGMs received from the full lesion model for different catheter orientations; Orthogonal orientation (ortho), parallel catheter orientation with distal to proximal propagation direction (para-dp) and parallel catheter orientation with proximal to distal catheter orientation (para-pd); Color code proximal EGM lesion transmurality: 0% (–), 28% (–), 50% (–),71% (–),92% (–),100% (–),110% (–); Color code distal UEGM: lesion transmuarlity 0% (–), 28% (–), 50% (–), 71% (–), 92% (–), 100% (–), 110% (–); Color code bipolar EGM: lesion transmurality: 0% (–), 28% (–), 50% (–),71% (–),92% (–),100% (–), 110% (–)

References

[1] V. Fuster and J. Hurst, *Hurst's the Heart, 12th Edition: Vol. 1.* McGraw-Hill Medical, 2008.

[2] H. Calkins, J. Brugada, D. L. Packer, R. Cappato, S.-A. Chen, H. J. G. Crijns, R. J. J. Damiano, D. W. Davies, D. E. Haines, M. Haissaguerre, Y. Iesaka, W. Jackman, P. Jais, H. Kottkamp, K. H. Kuck, B. D. Lindsay, F. E. Marchlinski, P. M. McCarthy, J. L. Mont, F. Morady, K. Nademanee, A. Natale, C. Pappone, E. Prystowsky, A. Raviele, J. N. Ruskin, R. J. Shemin, and Society of Thoracic Surgeons, "HRS/EHRA/ECAS expert consensus statement on catheter and surgical ablation of atrial fibrillation: recommendations for personnel, policy, procedures and follow-up. A report of the Heart Rhythm Society (HRS) Task Force on Catheter and Surgical Ablation of Atrial Fibrillation developed in partnership with the European Heart Rhythm Association (EHRA) and the European Cardiac Arrhythmia Society (ECAS); in collaboration with the American College of Cardiology (ACC), American Heart Association (AHA), and the Society of Thoracic Surgeons (STS). Endorsed and approved by the governing bodies of the American College of Cardiology, the American Heart Association, the European Cardiac Arrhythmia Society, the European Heart Rhythm Association, the Society of Thoracic Surgeons, and the Heart Rhythm Society," 2007.

[3] A. Natale and J. Jalife, *Atrial Fibrillation: From Bench to Bedside.* Contemporary Cardiology, Humana Press, 2008.

[4] C. Schmitt, I. Deisenhofer, and B. Zrenner, *Catheter Ablation of Cardiac Arrhythmias: A Practical Approach.* Medicine (Springer-11650; ZDB-2-SME), Steinkopff, 2006.

[5] K. Nademanee, J. McKenzie, E. Kosar, M. Schwab, B. Sunsaneewitayakul, T. Vasavakul, C. Khunnawat, and T. Ngarmukos, "A new approach for catheter ablation of atrial fibrillation: mapping of the electrophysiologic substrate," *J Am Coll Cardiol*, vol. 43, pp. 2044–53, Jun 2004.

[6] K. Otomo, K. Uno, H. Fujiwara, M. Isobe, and Y. Iesaka, "Local unipolar and bipolar electrogram criteria for evaluating the transmurality of atrial ablation lesions at different catheter orientations relative to the endocardial surface," *Heart Rhythm*, vol. 7, pp. 1291–300, Sep 2010.

[7] M. W. Keller, S. Schuler, A. Luik, G. Seemann, C. Schilling, C. Schmitt, and O. Dössel, "Comparison of simulated and clinical intracardiac electrograms," in *Proc IEEE EMBS*, (Osaka, Japan, 3 - 7 July), pp. 6858–6861, IEEE, 2013.

[8] S. Schuler, M. W. Keller, T. Oesterlein, G. Seemann, and O. Dössel, "Influence of Catheter Orientation, Tissue Thickness and Conduction Velocity on the Intracardiac Electrogram," in *Biomedical Engineering/Biomedizinische Technik*, vol. 58, 2013.

[9] M. W. Keller, S. Schuler, G. Seemann, and O. Dössel, "Differences in intracardiac signals on a realistic catheter geometry using mono and bidomain models," in *Computing in Cardiology*, vol. 39, (Krakow), pp. 305–308, 2012.

[10] M. W. Keller, A. Luik, M. Abady, G. Seemann, C. Schmitt, and O. Dössel, "Influence of three-dimensional fibrotic patterns on simulated intracardiac electrogram morphology," in *Computing in Cardiology*, 2013.

[11] M. W. Keller, S. Schuler, M. Wilhelms, G. Lenis, G. Seemann, C. Schmitt, O. Dössel, and A. Luik, "Characterization of Radiofrequency Ablation Lesion Development Based on Simulated and Mea-

sured Intracardiac Electrograms," *IEEE Transactions on Biomedical Engineering*, (published online ahead of print), 2014.

[12] J. Malmivuo and R. Plonsey, *Bioelectromagnetism: Principles and Applications of Bioelectric and Biomagnetic Fields.* Oxford University Press, 1995.

[13] L. Sherwood and C. L. (Firm), *Human Physiology: From Cells to Systems.* Brooks/Cole, Cengage Learning, 2011.

[14] S. Silbernagl and A. Despopoulos, *Color Atlas of Physiology.* Basic sciences, Thieme, 2009.

[15] J. J. Zipes, D. P., *Cardiac Electrophysiology From Cell to Bedside.* Saunders, Philadelphia, Pennsylvania, 4 ed., 2004.

[16] W. Ovalle, P. Nahirney, and F. Netter, *Netter's Essential Histology.* Netter Basic Science Series, Elsevier/Saunders, 2013.

[17] D. Sánchez-Quintana, J. A. Cabrera, V. Climent, J. Farré, M. C. d. Mendonça, and S. Y. Ho, "Anatomic relations between the esophagus and left atrium and relevance for ablation of atrial fibrillation," *Circulation*, vol. 112, pp. 1400–5, Sep 2005.

[18] L. S. Wann, A. B. Curtis, K. A. Ellenbogen, N. A. M. r. Estes, M. D. Ezekowitz, W. M. Jackman, C. T. January, J. E. Lowe, R. L. Page, D. J. Slotwiner, W. G. Stevenson, C. M. Tracy, V. Fuster, L. E. Ryden, D. S. Cannom, H. J. Crijns, J. L. Halperin, G. N. Kay, J.-Y. Le Heuzey, S. B. Olsson, E. N. Prystowsky, J. L. Tamargo, A. K. Jacobs, J. L. Anderson, N. Albert, M. A. Creager, S. M. Ettinger, R. A. Guyton, J. S. Hochman, F. G. Kushner, E. M. Ohman, and C. W. Yancy, "2011 ACCF/AHA/HRS focused update on the management of patients with atrial fibrillation (update on dabigatran). A report of the American College of Cardiology Foundation/American Heart Association Task Force on Practice Guidelines," *Heart Rhythm*, vol. 8, pp. e1–8, 2011.

[19] M. Haïssaguerre, P. Jaïs, D. C. Shah, A. Takahashi, M. Hocini, G. Quiniou, S. Garrigue, A. Le Mouroux, P. Le Métayer, and J. Clémenty, "Spontaneous initiation of atrial fibrillation by ectopic beats originating in the pulmonary veins," *N Engl J Med*, vol. 339, pp. 659–66, Sep 1998.

[20] C. Mahnkopf, T. J. Badger, N. S. Burgon, M. Daccarett, T. S. Haslam, C. T. Badger, C. J. McGann, N. Akoum, E. Kholmovski, R. S. Macleod, and N. F. Marrouche, "Evaluation of the left atrial substrate in patients with lone atrial fibrillation using delayed-enhanced MRI: implications for disease progression and response to catheter ablation," *Heart Rhythm*, vol. 7, pp. 1475–81, Oct 2010.

[21] S. Kostin, G. Klein, Z. Szalay, S. Hein, E. P. Bauer, and J. Schaper, "Structural correlate of atrial fibrillation in human patients," *Cardiovasc Res*, vol. 54, pp. 361–79, May 2002.

[22] V. Polyakova, S. Miyagawa, Z. Szalay, J. Risteli, and S. Kostin, "Atrial extracellular matrix remodelling in patients with atrial fibrillation," *Journal of Cellular and Molecular Medicine*, vol. 12, no. 1, pp. 189–208, 2008.

[23] J. Seitz, J. Horvilleur, J. Lacotte, D. O H-Ici, Y. Mouhoub, A. Maltret, F. Salerno, D. Myolette, M. Monchi, and J. Garot, "Correlation between AF Substrate Ablation Difficulty and Left Atrial Fibrosis Quantified by Delayed-Enhancement Cardiac Magnetic Resonance," *Pacing and Clinical Electrophysiology*, vol. 34, no. 10, pp. 1267–1277, 2011.

[24] M. Allessie, J. Ausma, and U. Schotten, "Electrical, contractile and structural remodeling during atrial fibrillation," *Cardiovasc Res*, vol. 54, pp. 230–46, May 2002.

[25] S. de Jong, T. A. van Veen, H. V. van Rijen, and J. M. de Bakker, "Fibrosis and cardiac arrhythmias," *Journal of cardiovascular pharmacology*, vol. 57, no. 6, pp. 630–638, 2011.

[26] T. P. Nguyen, Z. Qu, and J. N. Weiss, "Cardiac fibrosis and arrhythmogenesis: The road to repair is paved with perils," *J Mol Cell Cardiol*, Oct 2013.

[27] T. Kawara, R. Derksen, J. R. de Groot, R. Coronel, S. Tasseron, A. C. Linnenbank, R. N. Hauer, H. Kirkels, M. J. Janse, and J. M. de Bakker, "Activation delay after premature stimulation in chronically diseased

human myocardium relates to the architecture of interstitial fibrosis," *Circulation*, vol. 104, pp. 3069–75, Dec 2001.

[28] V. Jacquemet and C. S. Henriquez, "Genesis of complex fractionated atrial electrograms in zones of slow conduction: a computer model of microfibrosis," *Heart Rhythm*, vol. 6, pp. 803–10, Jun 2009.

[29] K. H. W. J. Ten Tusscher and A. V. Panfilov, "Influence of diffuse fibrosis on wave propagation in human ventricular tissue," *Europace*, vol. 9 Suppl 6, pp. vi38–45, Nov 2007.

[30] K. Nademanee, M. C. Schwab, E. M. Kosar, M. Karwecki, M. D. Moran, N. Visessook, A. D. Michael, and T. Ngarmukos, "Clinical outcomes of catheter substrate ablation for high-risk patients with atrial fibrillation," *J Am Coll Cardiol*, vol. 51, pp. 843–9, Feb 2008.

[31] F. Ouyang, R. Tilz, J. Chun, B. Schmidt, E. Wissner, T. Zerm, K. Neven, B. Köktürk, M. Konstantinidou, A. Metzner, A. Fuernkranz, and K.-H. Kuck, "Long-term results of catheter ablation in paroxysmal atrial fibrillation: lessons from a 5-year follow-up," *Circulation*, vol. 122, pp. 2368–77, Dec 2010.

[32] A. Verma, P. Sanders, J. Champagne, L. Macle, G. M. Nair, H. Calkins, and D. J. Wilber, "Selective Complex Fractionated Atrial Electrograms Targeting for Atrial Fibrillation Study (SELECT AF): A Multicenter, Randomized Trial," *Circ Arrhythm Electrophysiol*, vol. 7, pp. 55–62, Feb 2014.

[33] R. Weerasooriya, P. Khairy, J. Litalien, L. Macle, M. Hocini, F. Sacher, N. Lellouche, S. Knecht, M. Wright, I. Nault, S. Miyazaki, C. Scavee, J. Clementy, M. Haissaguerre, and P. Jais, "Catheter ablation for atrial fibrillation: are results maintained at 5 years of follow-up?," *J Am Coll Cardiol*, vol. 57, pp. 160–6, Jan 2011.

[34] E. Aliot, M. Haissaguerre, and W. M. Jackman, *Catheter Ablation of Atrial Fibrillation*. Wiley, 2011.

[35] F. H. Wittkampf, H. Nakagawa, W. S. Yamanashi, S. Imai, and W. M. Jackman, "Thermal latency in radiofrequency ablation," *Circulation*, vol. 93, pp. 1083–6, Mar 1996.

[36] S. Nath and D. E. Haines, "Biophysics and pathology of catheter energy delivery systems," *Prog Cardiovasc Dis*, vol. 37, pp. 185–204, 1995.

[37] M. Wood, S. Goldberg, M. Lau, A. Goel, D. Alexander, F. Han, and S. Feinstein, "Direct measurement of the lethal isotherm for radiofrequency ablation of myocardial tissue," *Circ Arrhythm Electrophysiol*, vol. 4, pp. 373–8, Jun 2011.

[38] D. E. Haines, "The biophysics of radiofrequency catheter ablation in the heart: the importance of temperature monitoring," *Pacing and Clinical Electrophysiology : PACE*, vol. 16, pp. 586–591, 1993.

[39] H. L. Estner, I. Deisenhofer, A. Luik, G. Ndrepepa, C. von Bary, B. Zrenner, and C. Schmitt, "Electrical isolation of pulmonary veins in patients with atrial fibrillation: reduction of fluoroscopy exposure and procedure duration by the use of a non-fluoroscopic navigation system (NavX)," *Europace : European Pacing, Arrhythmias, and Cardiac Electrophysiology : Journal of the Working Groups on Cardiac Pacing, Arrhythmias, and Cardiac Cellular Electrophysiology of the European Society of Cardiology*, vol. 8, pp. 583–587, 2006.

[40] S. Nath, J. P. DiMarco, and D. E. Haines, "Basic aspects of radiofrequency catheter ablation," *J Cardiovasc Electrophysiol*, vol. 5, pp. 863–876, 1994.

[41] D. E. Haines, *Catheter Ablation of Cardiac Arrhythmias: Basic Concepts and Clinical Applications*, vol. Third Edition, ch. Biophysics and pathophysiology of lesion formation by transcatheter radiofrequency ablation, pp. 20–34. Blackwell Futura, 2008.

[42] F. H. Wittkampf, R. N. Hauer, and E. O. Robles de Medina, "Control of radiofrequency lesion size by power regulation," *Circulation*, vol. 80, pp. 962–968, 1989.

[43] D. E. Haines, "Determinants of Lesion Size During Radiofrequency Catheter Ablation: The Role of Electrode-Tissue Contact Pressure and Duration of Energy Delivery," *Journal of Cardiovascular Electrophysiology*, vol. 2, pp. 509–515, 1991.

[44] G. Ndrepepa and H. Estner, *Catheter Ablation of Cardiac Arrhythmias: A Practical Approach*, vol. Catheter Ablation of Cardiac Arrhythmias, ch. Ablation of cardiac arrhythmias: energy sources and mechanisms of lesion formation, pp. 35–53. Steinkopff, 2006.

[45] M. K. Jain and P. D. Wolf, "A three-dimensional finite element model of radiofrequency ablation with blood flow and its experimental validation," *Ann Biomed Eng*, vol. 28, pp. 1075–84, Sep 2000.

[46] Y.-C. Lai, Y. B. Choy, D. Haemmerich, V. R. Vorperian, and J. G. Webster, "Lesion size estimator of cardiac radiofrequency ablation at different common locations with different tip temperatures," *IEEE Trans Biomed Eng*, vol. 51, pp. 1859–64, Oct 2004.

[47] J. P. Saul, J. E. Hulse, J. Papagiannis, R. Van Praagh, and E. P. Walsh, "Late enlargement of radiofrequency lesions in infant lambs. Implications for ablation procedures in small children," *Circulation*, vol. 90, pp. 492–9, Jul 1994.

[48] D. E. Haines, *Catheter Ablation of Arrhythmias, 2nd Edition*, ch. Pathophysiology of radiofrequency lesion formation and the role of new energy modalities, pp. 67–88. NY: Futura Publishing, Inc, 2002.

[49] S. Nath, C. r. Lynch, J. G. Whayne, and D. E. Haines, "Cellular electrophysiological effects of hyperthermia on isolated guinea pig papillary muscle. Implications for catheter ablation," *Circulation*, vol. 88, pp. 1826–1831, 1993.

[50] T. A. Simmers, J. M. De Bakker, F. H. Wittkampf, and R. N. Hauer, "Effects of heating on impulse propagation in superfused canine myocardium," *J Am Coll Cardiol*, vol. 25, pp. 1457–64, May 1995.

[51] T. A. Simmers, J. M. de Bakker, F. H. Wittkampf, and R. N. Hauer, "Effects of heating with radiofrequency power on myocardial impulse conduction: is radiofrequency ablation exclusively thermally mediated?," *J Cardiovasc Electrophysiol*, vol. 7, pp. 243–7, Mar 1996.

[52] M. A. Wood and I. A. Fuller, "Acute and chronic electrophysiologic changes surrounding radiofrequency lesions," *J Cardiovasc Electrophysiol*, vol. 13, pp. 56–61, 2002.

[53] Y. Z. Ge, P. Z. Shao, J. Goldberger, and A. Kadish, "Cellular electrophysiological changes induced in vitro by radiofrequency current: comparison with electrical ablation," *PACE*, vol. 18, pp. 323–333, 1995.

[54] C.-C. Wu, R. W. Fasciano, H. Calkins, and L. Tung, "Sequential change in action potential of rabbit epicardium during and following radiofrequency ablation," *Journal of Cardiovascular Electrophysiology*, vol. 10, pp. 1252–1261, 1999.

[55] S. Nath, J. G. Whayne, S. Kaul, N. C. Goodman, A. R. Jayaweera, and D. E. Haines, "Effects of radiofrequency catheter ablation on regional myocardial blood flow. Possible mechanism for late electrophysiological outcome," *Circulation*, vol. 89, pp. 2667–2672, 1994.

[56] T. H. t. Everett, S. Nath, C. r. Lynch, J. M. Beach, J. G. Whayne, and D. E. Haines, "Role of calcium in acute hyperthermic myocardial injury," *Journal of Cardiovascular Electrophysiology*, vol. 12, pp. 563–569, 2001.

[57] P. N. Yi, C. S. Chang, M. Tallen, W. Bayer, and S. Ball, "Hyperthermia-induced intracellular ionic level changes in tumor cells," *Radiation Research*, vol. 93, pp. 534–544, 1983.

[58] J. Goldberger and J. Ng, *Practical Signal and Image Processing in Clinical Cardiology*. Springer, 2010.

[59] J. M. T. de Bakker and F. H. M. Wittkampf, "The pathophysiologic basis of fractionated and complex electrograms and the impact of recording techniques on their detection and interpretation," *Circ Arrhythm Electrophysiol*, vol. 3, pp. 204–13, Apr 2010.

[60] G. Sverre and G. Martinsen Ørjan, "Bioimpedance and Bioelectricity Basics," 2008.

[61] R. H. Milad El Haddad, "Algorithmic detection of the beginning and end of bipolar electrograms: Implications for novel methods to assess local activation time during atrial tachycardia," *Biomedical Signal Processing and Control*, 2012.

[62] C. Schilling, *Analysis of Atrial Electrograms (Karlsruhe transactions on biomedical engineering) (Volume 17)*. KIT Scientific Publishing, 2012.

[63] M. P. Nguyen, C. Schilling, and O. Dössel, "A new approach for automated location of active segments in intracardiac electrograms," in *IFMBE Proceedings World Congress on Medical Physics and Biomedical Engineering*, vol. 25/4, pp. 763–766, 2009.

[64] J. F. Kaiser, "On a simple algorithm to calculate the energy'of a signal," in *Acoustics, Speech, and Signal Processing, 1990. ICASSP 90., 1990 International Conference on*, pp. 381–384, IEEE, 1990.

[65] M. Aubreville, C. Schilling, A. Luik, F. M. Weber, C. Schmitt, and O. Dössel, "Detection of periodicity in complex fractionated atrial electrograms (CFAEs)," in *IFMBE Proceedings World Congress on Medical Physics and Biomedical Engineering*, vol. 25/4, (Munich, Germany), 2009.

[66] A. S. Jadidi, H. Cochet, A. J. Shah, S. J. Kim, E. Duncan, S. Miyazaki, M. Sermesant, H. Lehrmann, M. Lederlin, N. Linton, A. Forclaz, I. Nault, L. Rivard, M. Wright, X. Liu, D. Scherr, S. B. Wilton, L. Roten, P. Pascale, N. Derval, F. Sacher, S. Knecht, C. Keyl, M. Hocini, M. Montaudon, F. Laurent, M. Haïssaguerre, and P. Jaïs, "Inverse relationship between fractionated electrograms and atrial fibrosis in persistent atrial fibrillation: combined magnetic resonance imaging and high-density mapping," *J Am Coll Cardiol*, vol. 62, pp. 802–12, Aug 2013.

[67] F. B. Sachse, N. S. Torres, E. Savio-Galimberti, T. Aiba, D. A. Kass, G. F. Tomaselli, and J. H. Bridge, "Subcellular structures and function of myocytes impaired during heart failure are restored by cardiac resynchronization therapy," *Circ Res*, vol. 110, pp. 588–97, Feb 2012.

[68] S. Rohr and J. P. Kucera, "Optical Recording System Based on a Fiber Optic Image Conduit: Assessment of Microscopic Activation Patterns in Cardiac Tissue," *Biophysical Journal*, vol. 75, no. 2, pp. 1062 – 1075, 1998.

[69] R. Mandapati, A. Skanes, J. Chen, O. Berenfeld, and J. Jalife, "Stable microreentrant sources as a mechanism of atrial fibrillation in the isolated sheep heart," *Circulation*, vol. 101, no. 2, pp. 194–199, 2000.

[70] A. V. Glukhov, V. V. Fedorov, Q. Lou, V. K. Ravikumar, P. W. Kalish, R. B. Schuessler, N. Moazami, and I. R. Efimov, "Transmural dispersion of repolarization in failing and nonfailing human ventricle," *Circulation research*, vol. 106, no. 5, pp. 981–991, 2010.

[71] I. R. Efimov, V. P. Nikolski, and G. Salama, "Optical imaging of the heart," *Circ Res*, vol. 95, pp. 21–33, Jul 2004.

[72] A. Nygren, A. Lomax, and W. Giles, "Optical Mapping System for Recording Action Potential Durations in Adult Mouse Left and Right Atrium," in *Engineering in Medicine and Biology Society, 2004. IEMBS '04. 26th Annual International Conference of the IEEE*, vol. 2, pp. 3576 – 3577, sept. 2004.

[73] M. Attin and W. T. Clusin, "Basic concepts of optical mapping techniques in cardiac electrophysiology," *Biol Res Nurs*, vol. 11, pp. 195–207, Oct 2009.

[74] D. B. Cowan, B. Sill, and P. E. Hammer, "Optical Mapping of Langendorff-perfused Rat Hearts," *J Vis Exp*, 08 2009.

[75] A. Matiukas, A. Pertsov, P. Kothari, A. Cram, and E. Tolkacheva, "Optical mapping of electrical heterogeneities in the heart during global ischemia," in *Proc IEEE EMBS*, pp. 6321 –6324, sep. 2009.

[76] D. S. Rosenbaum, ed., *Optical mapping of cardiac excitation and arrhythmias*. Armonk, NY: Futura, 2001. Includes bibliographical references and index.

[77] B. Kuhn and P. Fromherz, "Anellated Hemicyanine Dyes in a Neuron Membrane:âĂŽÄâ Molecular Stark Effect and Optical Voltage Recording," *The Journal of Physical Chemistry B*, vol. 107, no. 31, pp. 7903–7913, 2003.

[78] E. Fluhler, V. G. Burnham, and L. M. Loew, "Spectra, membrane binding, and potentiometric responses of new charge shift probes," *Biochemistry*, vol. 24, no. 21, pp. 5749–5755, 1985.

[79] M. Canepari and D. Zecevic, *Membrane potential imaging in the nervous system.* Springer, 2010.

[80] R. D. Walton, B. G. Mitrea, A. M. Pertsov, and O. Bernus, "A novel near-infrared voltage-sensitive dye reveals the action potential wavefront orientation at increased depths of cardiac tissue," *Conf Proc IEEE Eng Med Biol Soc*, vol. 2009, pp. 4523–6, 2009.

[81] P. Fromherz, G. Hübener, B. Kuhn, and M. J. Hinner, "ANNINE-6plus, a voltage-sensitive dye with good solubility, strong membrane binding and high sensitivity," *Eur Biophys J*, vol. 37, pp. 509–14, Apr 2008.

[82] W. Baxter, J. Davidenko, L. Loew, J. Wuskell, and J. Jalife, "Technical features of a CCD video camera system to record cardiac fluorescence data," *Annals of Biomedical Engineering*, vol. 25, pp. 713–725, 1997. 10.1007/BF02684848.

[83] J. I. Laughner, F. S. Ng, M. S. Sulkin, R. M. Arthur, and I. R. Efimov, "Processing and analysis of cardiac optical mapping data obtained with potentiometric dyes," *Am J Physiol Heart Circ Physiol*, vol. 303, pp. H753–65, Oct 2012.

[84] "Practical Methods in Cardiovascular Research," 2005. In: Springer-Online.

[85] E. H. Ratzlaff and A. Grinvald, "A tandem-lens epifluorescence macroscope: Hundred-fold brightness advantage for wide-field imaging," *Journal of Neuroscience Methods*, vol. 36, no. 2-3, pp. 127 – 137, 1991.

[86] S. D. Girouard, K. R. Laurita, and D. S. Rosenbaum, "Unique properties of cardiac action potentials recorded with voltage-sensitive dyes," *J Cardiovasc Electrophysiol*, vol. 7, pp. 1024–38, Nov 1996.

[87] J. Abernethy, "The boxcar detector," *Wireless World*, vol. 76, 1970.

[88] S.-F. Lin, R. A. Abbas, and J. P. Wikswo, "High-resolution high-speed synchronous epifluorescence imaging of cardiac activation," *Review of scientific instruments*, vol. 68, no. 1, pp. 213–217, 1997.

[89] J. Thiele, *Optische und mechanische Messungen von elektrophysiologischen Vorgängen im Myokardgewebe*. PhD thesis, Universitaet Karlsruhe (TH), Universitätsverlag Karlsruhe, 2008.

[90] F. Bretschneider, *Introduction to electrophysiological methods and instrumentation*. Amsterdam: Elsevier/Academic Press, 1st ed. ed., 2006.

[91] O. G. Martinsen and S. Grimnes, *Bioimpedance and bioelectricity basics*. Academic press, 2000.

[92] W. Franks, I. Schenker, P. Schmutz, and A. Hierlemann, "Impedance characterization and modeling of electrodes for biomedical applications," *IEEE Trans Biomed Eng*, vol. 52, pp. 1295–302, Jul 2005.

[93] E. McAdams, A. Lackermeier, J. McLaughlin, D. Macken, and J. Jossinet, "The linear and non-linear electrical properties of the electrode-electrolyte interface," *Biosensors and Bioelectronics*, vol. 10, no. 1, pp. 67–74, 1995.

[94] R. Arnold, T. Wiener, T. Thurner, and E. Hofer, "A Novel Electrophysiological Measurement System to Study Rapidly Paced Animal Hearts," in *4th European Conference of the Interna-tional Federation for Medical and Biological Engineering* (J. Sloten, P. Verdonck, M. Nyssen, and J. Haueisen, eds.), vol. 22 of *IFMBE Proceedings*, pp. 1145–1148, Springer Berlin Heidelberg, 2009. 10.1007/978-3-540-89208-3_274.

[95] F. B. Sachse, B. W. Steadman, J. H. B. B Bridge, B. B. Punske, and B. Taccardi, "Conduction velocity in myocardium modulated by strain: measurement instrumentation and initial results," *Conf Proc IEEE Eng Med Biol Soc*, vol. 5, pp. 3593–6, 2004.

[96] R. Arnold, *Beat-to-Beat Behavior of Atrial Activation Sequences under Impaired Conduction Conditions*. PhD thesis, Medical Univerisity of Graz, 2013.

[97] R. Arnold, T. Wiener, D. Sanchez-Quintana, and E. Hofer, "Topology and conduction in the inferior right atrial isthmus measured in rabbit hearts," *Conf Proc IEEE Eng Med Biol Soc*, vol. 2011, pp. 247–50, 2011.

[98] L. S. Matsubara, B. B. Matsubara, M. P. Okoshi, A. C. Cicogna, and J. S. Janicki, "Alterations in myocardial collagen content affect rat papillary muscle function," *Am J Physiol Heart Circ Physiol*, vol. 279, pp. H1534–9, Oct 2000.

[99] J. R. Silva, H. Pan, D. Wu, A. Nekouzadeh, K. F. Decker, J. Cui, N. A. Baker, D. Sept, and Y. Rudy, "A multiscale model linking ion-channel molecular dynamics and electrostatics to the cardiac action potential," *Proc Natl Acad Sci U S A*, vol. 106, pp. 11102–6, Jul 2009.

[100] D. B. Geselowitz, R. C. Barr, M. S. Spach, and W. T. Miller, 3rd, "The impact of adjacent isotropic fluids on electrograms from anisotropic cardiac muscle. A modeling study," *Circ Res*, vol. 51, pp. 602–13, Nov 1982.

[101] D. B. Geselowitz and T. W. Miller, "A bidomain model for anisotropic cardiac muscle," *Annals of Biomedical Engineering*, vol. 11, pp. 191–206, 1983.

[102] W. T. Miller and D. B. Geselowitz, "Simulation studies of the electrocardiogram. I. The normal heart," *Circ Res*, vol. 43, pp. 301–15, Aug 1978.

[103] M. Courtemanche, R. J. Ramirez, and S. Nattel, "Ionic mechanisms underlying human atrial action potential properties: Insights from a mathematical model," *Am. J. Physiol.*, vol. 275, pp. H301–H321, 1998.

[104] K. H. W. J. ten Tusscher and A. V. Panfilov, "Alternans and spiral breakup in a human ventricular tissue model," *Am J Physiol*, vol. 291, pp. H1088–100, 2006.

[105] M. Wilhelms, O. Dössel, and G. Seemann, "Comparing Simulated Electrocardiograms of Different Stages of Acute Cardiac Ischemia," in *FIMH 2011, LNCS*, vol. 6666, pp. 11–19, 2011.

[106] A. L. Hodgkin and A. F. Huxley, "A quantitative description of membrane current and its application to conduction and excitation in nerve," *Journal of Physiology*, vol. 117, pp. 500–544, 1952.

[107] G. Seemann, *Modeling of electrophysiology and tension development in the human heart*. PhD thesis, Universitaet Karlsruhe (TH), Universitätsverlag Karlsruhe, Institut für Biomedizinische Technik, 2005.

[108] M. Wilhelms, H. Hettmann, M. M. C. Maleckar, J. T. Koivumäki, O. Dössel, and G. Seemann, "Benchmarking electrophysiological models of human atrial myocytes," *Frontiers in Physiology*, vol. 3, pp. 1–16, 2013.

[109] M. Wilhelms, *Multiscale Modeling of Cardiac Electrophysiology: Adaptation to Atrial and Ventricular Rhythm Disorders and Pharmacological Treatment*. PhD thesis, Karlsruhe Institute of Technology (KIT), 2013.

[110] K. H. W. J. t. Tusscher, D. Noble, P. J. Noble, and A. V. Panfilov, "A model for human ventricular tissue," *Am. J. Physiol.*, vol. 286, pp. H1573–H1589, 2004.

[111] D. L. Weiss, M. Ifland, F. B. Sachse, G. Seemann, and O. Dössel, "Modeling of cardiac ischemia in human myocytes and tissue including spatiotemporal electrophysiological variations," *Biomed Tech (Berl)*, vol. 54, pp. 107–125, 2009.

[112] M. S. Spach, J. F. Heidlage, P. C. Dolber, and R. C. Barr, "Mechanism of origin of conduction disturbances in aging human atrial bundles: experimental and model study," *Heart Rhythm*, vol. 4, pp. 175–85, Feb 2007.

[113] C. S. Henriquez, "Simulating the electrical behavior of cardiac tissue using the bidomain model," *Crit Rev Biomed Eng*, vol. 21, no. 1, pp. 1–77, 1993.

[114] G. Seemann, F. B. Sachse, M. Karl, D. L. Weiss, V. Heuveline, and O. Dössel, "Framework for modular, flexible and efficient solving the cardiac bidomain equation using PETSc," *Progr. Industr. Math.*, vol. 15, pp. 363–369, 2010.

[115] V. Jacquemet, L. Kappenberger, and C. S. Henriquez, "Modeling atrial arrhythmias: impact on clinical diagnosis and therapies," *IEEE Rev Biomed Eng*, vol. 1, pp. 94–114, 2008.

[116] F. Campos, T. Wiener, A. Prassl, R. Weber Dos Santos, D. Sanchez-Quintana, H. Ahammer, G. Plank, and E. Hofer, "Electro-Anatomical Characterization of Atrial Microfibrosis in a Histologically Detailed Computer Model," *IEEE Trans Biomed Eng*, Apr 2013.

[117] V. Jacquemet, N. Virag, Z. Ihara, L. Dang, O. Blanc, S. Zozor, J.-M. Vesin, L. Kappenberger, and C. Henriquez, "Study of unipolar electrogram morphology in a computer model of atrial fibrillation," *J Cardiovasc Electrophysiol*, vol. 14, pp. S172–9, 2003.

[118] M. Sadiku, *Numerical Techniques in Electromagnetics, Second Edition*. Taylor & Francis, 2000.

[119] MATLAB, *version 8.2.0 (R2013b)*. Natick, Massachusetts: The MathWorks Inc., 2010.

[120] E. T. Castellana and P. S. Cremer, "Imaging large arrays of supported lipid bilayers with a macroscope," *Biointerphases*, vol. 2, pp. 57–63, Jun 2007.

[121] T. Lewalter, C. Weiss, S. Spencker, W. Jung, W. Haverkamp, S. Willems, T. Deneke, J. Kautzner, M. Wiedemann, J. Siebels, H. F. Pitschner, E. Hoffmann, G. Hindricks, M. Zabel, E. Vester, H. Schwacke, E. Mittmann-Braun, L. Lickfett, S. Hoffmeister, J. Proff, C. Mewis, W. Bauer, and AURUM 8 Study Investigators, "Gold vs. platinum-iridium tip catheter for cavotricuspid isthmus ablation: the AURUM 8 study," *Europace*, vol. 13, pp. 102–8, Jan 2011.

[122] T. Söderström, *System identification*. New York: Prentice Hall, 1989.

[123] P. Welch, "The use of fast Fourier transform for the estimation of power spectra: a method based on time averaging over short, modified periodograms," *Audio and Electroacoustics, IEEE Transactions on*, vol. 15, no. 2, pp. 70–73, 1967.

[124] T. Deneke, K. Khargi, K.-M. Müller, B. Lemke, A. Mügge, A. Laczkovics, A. E. Becker, and P. H. Grewe, "Histopathology of intraoperatively induced linear radiofrequency ablation lesions in patients with chronic atrial fibrillation," *Eur Heart J*, vol. 26, pp. 1797–803, Sep 2005.

[125] P. G. Platonov, V. Ivanov, S. Y. Ho, and L. Mitronova, "Left Atrial Posterior Wall Thickness in Patients with and without Atrial Fibrillation: Data from 298 Consecutive Autopsies," *Journal of Cardiovascular Electrophysiology*, vol. 19, no. 7, pp. 689–692, 2008.

[126] B. Hall, V. Jeevanantham, R. Simon, J. Filippone, G. Vorobiof, and J. Daubert, "Variation in left atrial transmural wall thickness at sites commonly targeted for ablation of atrial fibrillation," *Journal of Interventional Cardiac Electrophysiology*, vol. 17, no. 2, pp. 127–132, 2006.

[127] S. Gabriel, R. W. Lau, and C. Gabriel, "The dielectric properties of biological tissues: III. Parametric models for the dielectric spectrum of tissues," *Physics in Medicine and Biology*, vol. 41, pp. 2271–2293, 1996.

[128] B. C. Schwab, G. Seemann, R. A. Lasher, N. S. Torres, E. M. Wulfers, M. Arp, E. D. Carruth, J. H. B. Bridge, and F. B. Sachse, "Quantitative analysis of cardiac tissue including fibroblasts using three-dimensional confocal microscopy and image reconstruction: Towards a basis for electrophysiological modeling," *IEEE Transactions on Medical Imaging*, vol. 32, pp. 862–872, 2013.

[129] K. Visser, "Electric conductivity of stationary and flowing human blood at low frequencies," in *Engineering in Medicine and Biology Society, 1989. Images of the Twenty-First Century., Proceedings of the Annual International Conference of the IEEE Engineering in*, pp. 1540–1542 vol.5, 1989.

[130] G. White and S. Woods, "Thermal and Electrical Conductivity of Rhodium, Iridium, and Platinum," *Canadian Journal of Physics*, vol. 35, no. 3, pp. 248–257, 1957.

[131] F. Xiang, H. Wang, and X. Yao, "Preparation and dielectric properties of bismuth-based dielectric/PTFE microwave composites," *Journal of the European Ceramic Society*, vol. 26, no. 10âĂŽÄì11, pp. 1999 – 2002, 2006. Papers Presented at the Third International Conference on Microwave Materials and their Applications - {MMA2004} Inuyama,

Japan The Third International Conference on Microwave Materials and their Applications - {MMA2004}.

[132] F. M. Weber, C. Schilling, G. Seemann, A. Luik, C. Schmitt, C. Lorenz, and O. Dössel, "Wave-direction and conduction-velocity analysis from intracardiac electrograms–a single-shot technique," *IEEE Transactions on Biomedical Engineering*, vol. 57, pp. 2394–2401, 2010.

[133] C. Lima, R. De Oliveira, S. Figueiro, C. Wehmann, J. Góes, and A. Sombra, "DC conductivity and dielectric permittivity of collagen–chitosan films," *Materials chemistry and physics*, vol. 99, no. 2, pp. 284–288, 2006.

[134] D. E. Haines and D. D. Watson, "Tissue heating during radiofrequency catheter ablation: a thermodynamic model and observations in isolated perfused and superfused canine right ventricular free wall," *Pacing Clin Electrophysiol*, vol. 12, pp. 962–76, Jun 1989.

[135] Y.-C. Lai, J. Webster, D. Beebe, Y. Hu, H. Ploeg, W. Tompkins, J. Will, and F. O. Defense, *Lesion Size Estimator for Cardiac Radio-frequency Ablation*. PhD thesis, Citeseer, 2009.

[136] S. Tungjitkusolmun, E. J. Woo, H. Cao, J. Z. Tsai, V. R. Vorperian, and J. G. Webster, "Thermal–electrical finite element modelling for radio frequency cardiac ablation: effects of changes in myocardial proper-ties," *Med Biol Eng Comput*, vol. 38, pp. 562–568, 2000.

[137] D. Haemmerich, "Biophysics of radiofrequency ablation," *Crit Rev Biomed Eng*, vol. 38, pp. 53–63, 2010.

[138] J. H. Dumas Iii, H. D. Himel Iv, A. C. Kiser, S. R. Quint, and S. B. Knisley, "Myocardial electrical impedance as a predictor of the quality of RF-induced linear lesions," *Physiol Meas*, vol. 29, pp. 1195–207, Oct 2008.

[139] T. A. Simmers, J. M. De Bakker, F. H. Wittkampf, and R. N. Hauer, "Effects of heating on impulse propagation in superfused canine my-ocardium," *J Am Coll Cardiol*, vol. 25, pp. 1457–64, May 1995.

[140] G. Fischer, B. Tilg, R. Modre, F. Hanser, B. Messnarz, and P. Wach, "On modeling the Wilson terminal in the boundary and finite element method," *Biomedical Engineering, IEEE Transactions on*, vol. 49, pp. 217–224, March 2002.

[141] Y. Bourgault, Y. Coudière, and C. Pierre, "Existence and uniqueness of the solution for the bidomain model used in cardiac electrophysiology," *Nonlinear Analysis: Real World Applications*, vol. 10, no. 1, pp. 458 – 482, 2009.

[142] U. B. Tedrow and W. G. Stevenson, "Recording and interpreting unipolar electrograms to guide catheter ablation," *Heart Rhythm*, vol. 8, pp. 791–6, May 2011.

[143] U. Richter, L. Faes, A. Cristoforetti, M. Masè, F. Ravelli, M. Stridh, and L. Sörnmo, "A novel approach to propagation pattern analysis in intracardiac atrial fibrillation signals," *Ann Biomed Eng*, vol. 39, pp. 310–23, Jan 2011.

[144] X. Zhu and D. Wei, "Computer simulation of intracardiac potential with whole-heart model," *Int J Bioinform Res Appl*, vol. 3, no. 1, pp. 100–22, 2007.

[145] C. Tobón, C. A. Ruiz-Villa, E. Heidenreich, L. Romero, F. Hornero, and J. Saiz, "A three-dimensional human atrial model with fiber orientation. Electrograms and arrhythmic activation patterns relationship," *PLoS One*, vol. 8, no. 2, p. e50883, 2013.

[146] J. V. Tranquillo, M. R. Franz, B. C. Knollmann, A. P. Henriquez, D. A. Taylor, and C. S. Henriquez, "Genesis of the monophasic action potential: role of interstitial resistance and boundary gradients," *Am J Physiol Heart Circ Physiol*, vol. 286, pp. H1370–81, Apr 2004.

[147] P. C. Franzone, L. F. Pavarino, S. Scacchi, and B. Taccardi, "Monophasic action potentials generated by bidomain modeling as a tool for detecting cardiac repolarization times," *Am J Physiol Heart Circ Physiol*, vol. 293, pp. H2771–85, Nov 2007.

[148] W. Mueller, H. Windisch, and H. Tritthart, "Fast optical monitoring of microscopic excitation patterns in cardiac muscle," *Biophysical Journal*, vol. 56, no. 3, pp. 623–629, 1989.

[149] V. V. Fedorov, W. J. Hucker, H. Dobrzynski, L. V. Rosenshtraukh, and I. R. Efimov, "Postganglionic nerve stimulation induces temporal inhibition of excitability in rabbit sinoatrial node," *Am J Physiol Heart Circ Physiol*, vol. 291, pp. H612–23, Aug 2006.

[150] T. Sakai, "Optical mapping of the spread of excitation in the isolated rat atrium during tachycardia-like excitation," *Pflügers Archiv European Journal of Physiology*, vol. 447, pp. 280–288, 2003. 10.1007/s00424-003-1185-x.

[151] V. Fedorov, I. Lozinsky, E. Sosunov, E. Anyukhovsky, M. Rosen, C. Balke, and I. Efimov, "Application of blebbistatin as an excitation-contraction uncoupler for electrophysiologic study of rat and rabbit hearts," *Heart Rhythm : the Official Journal of the Heart Rhythm Society*, vol. 4, pp. 619–626, 2007.

[152] J. Liau, J. Dumas, D. Janks, B. J. Roth, and S. B. Knisley, "Cardiac optical mapping under a translucent stimulation electrode," *Ann Biomed Eng*, vol. 32, pp. 1202–10, Sep 2004.

[153] H. D. Himel, 4th and S. B. Knisley, "Comparison of optical and electrical mapping of fibrillation," *Physiol Meas*, vol. 28, pp. 707–19, Jun 2007.

[154] S. B. Knisley and M. R. Neuman, "Simultaneous Electrical and Optical Mapping in Rabbit Hearts," *Annals of Biomedical Engineering*, vol. 31, pp. 32–41, 2003. 10.1114/1.1535413.

[155] S. B. Knisley and A. E. Pollard, "Use of translucent indium tin oxide to measure stimulatory effects of a passive conductor during field stimulation of rabbit hearts," *Am J Physiol Heart Circ Physiol*, vol. 289, pp. H1137–46, Sep 2005.

[156] S. B. Knisley, H. D. Himel, and J. H. Dumas, "Simultaneous Optical and Electrical Recordings," in *Cardiac Bioelectric Therapy* (I. R. Efimov, M. W. Kroll, and P. J. Tchou, eds.), pp. 357–380, Springer US, 2009. 10.1007/978-0-387-79403-7_14.

[157] M. S. Spach and J. M. Kootsey, "Relating the Sodium Current and Conductance to the Shape of Transmembrane and Extracellular Potentials by Simulation: Effects of Propagation Boundaries," *Biomedical Engineering, IEEE Transactions on*, vol. BME-32, pp. 743–755, Oct 1985.

[158] M. S. Spach, R. C. Barr, E. A. Johnson, and J. M. Kootsey, "Cardiac extracellular potentials: analysis of complex wave forms about the Purkinje networks in dogs," *Circulation Research*, vol. 33, pp. 465–473, 1973.

[159] M. Spach and P. C. Dolber, "Relating extracellular potentials and their derivatives to anisotropic propagation at a microscopic level in the human cardiac muscle. Evidence for electrical uncoupling of side-to-side fiber connections with increasing age," *Circulation Research*, vol. 58, pp. 356–371, 1986.

[160] T. Wiener, F. O. Campos, G. Plank, and E. Hofer, "Decomposition of fractionated local electrograms using an analytic signal model based on sigmoid functions," *Biomed Tech (Berl)*, vol. 0, pp. 1–12, Oct 2012.

[161] M. J. Bishop, E. Vigmond, and G. Plank, "Cardiac bidomain bath-loading effects during arrhythmias: interaction with anatomical heterogeneity," *Biophys J*, vol. 101, pp. 2871–81, Dec 2011.

[162] J. C. Eason and R. A. Malkin, "A simulation study evaluating the performance of high-density electrode arrays on myocardial tissue," *IEEE Trans Biomed Eng*, vol. 47, pp. 893–901, Jul 2000.

[163] U. Schotten, S. Verheule, P. Kirchhof, and A. Goette, "Pathophysiological mechanisms of atrial fibrillation: a translational appraisal," *Physiol Rev*, vol. 91, pp. 265–325, Jan 2011.

[164] N. Akoum, M. Daccarett, C. Mcgann, N. Segerson, G. Vergara, S. Kuppahally, T. Badger, N. Burgon, T. Haslam, E. Kholmovski, R. Macleod, and N. Marrouche, "Atrial Fibrosis Helps Select the Appropriate Patient and Strategy in Catheter Ablation of Atrial Fibrillation: A DE-MRI Guided Approach," *Journal of Cardiovascular Electrophysiology*, vol. 22, no. 1, pp. 16–22, 2011.

[165] K. S. McDowell, F. Vadakkumpadan, R. Blake, J. Blauer, G. Plank, R. S. MacLeod, and N. A. Trayanova, "Methodology for patient-specific modeling of atrial fibrosis as a substrate for atrial fibrillation," *Journal of Electrocardiology*, vol. 45, no. 6, pp. 640 – 645, 2012.

[166] M. W. Krueger, K. S. Rhode, M. D. O'Neill, C. A. Rinaldi, J. Gill, R. Razavi, G. Seemann, and O. Doessel, "Patient-specific modeling of atrial fibrosis increases the accuracy of sinus rhythm simulations and may explain maintenance of atrial fibrillation," *J Electrocardiol*, Nov 2013.

[167] C. M. Costa, F. O. Campos, A. J. Prassl, R. W. Dos Santos, D. Sanchez-Quintana, H. Ahammer, E. Hofer, and G. Plank, "An efficient finite element approach for modeling fibrotic clefts in the heart," *IEEE Trans Biomed Eng*, vol. 61, pp. 900–10, Mar 2014.

[168] A. Bortone, A. Appetiti, A. Bouzeman, E. Maupas, V. Ciobotaru, J.-M. Boulenc, P. Pujadas-Berthault, and P. Rioux, "Unipolar Signal Modification as a Guide for Lesion Creation during Radiofrequency Application in the Left Atrium: A Prospective Study in Humans in the Setting of Paroxysmal Atrial Fibrillation Catheter Ablation," *Circ Arrhythm Electrophysiol*, Oct 2013.

[169] M. Reumann, J. Bohnert, G. Seemann, B. Osswald, and O. Dössel, "Preventive ablation strategies in a biophysical model of atrial fibrillation based on realistic anatomical data," *IEEE Trans. Biomed. Eng.*, vol. 55, pp. 399–406, 2008.

[170] M. Rotter, L. Dang, V. Jacquemet, N. Virag, L. Kappenberger, and M. Haissaguerre, "Impact of varying ablation patterns in a simulation model of persistent atrial fibrillation," *PACE*, vol. 30, pp. 314–321, 2007.

[171] L. Dang, N. Virag, Z. Ihara, V. Jacquemet, J. M. Vesin, J. Schlaepfer, P. Ruchat, and L. Kappenberger, "Evaluation of ablation patterns using a biophysical model of atrial fibrillation," *Ann Biomed Eng*, vol. 33, pp. 465–474, 2005.

[172] M. W. Krueger, G. Seemann, K. Rhode, D. U. J. Keller, C. Schilling, A. Arujuna, J. Gill, M. D. O'Neill, R. Razavi, and O. Dössel, "Personalization of Atrial Anatomy and Elelectophysiology as a Basis for Clinical Modeling of Radio-Frequency-Ablation of Atrial Fibrillation," *IEEE Trans. Med. Imag.*, vol. 32, pp. 73–84, 2013.

[173] M. W. Krüger, *Personalized Multi-Scale Modeling of the Atria : Heterogeneities, Fiber Architecture, Hemodialysis and Ablation Therapy*. PhD thesis, Karlsruhe, KIT Scientific Publishing, 2013. Zugl.: Karlsruhe, KIT, Diss., 2012.

[174] F. O. Campos, A. J. Prassl, G. Seemann, R. Weber dos Santos, G. Plank, and E. Hofer, "Influence of ischemic core muscle fibers on surface depolarization potentials in superfused cardiac tissue preparations: a simulation study," *Med Biol Eng Comput*, vol. 50, pp. 461–72, May 2012.

[175] A. Barnette, P. Bayly, S. Zhang, G. P. Walcott, R. Ideker, and W. Smith, "Estimation of 3-D conduction velocity vector fields from cardiac mapping data," *Biomedical Engineering, IEEE Transactions on*, vol. 47, pp. 1027–1035, Aug 2000.

[176] P. V. Bayly, B. H. KenKnight, J. M. Rogers, R. E. Hillsley, R. E. Ideker, and W. M. Smith, "Estimation of conduction velocity vector fields from epicardial mapping data," *IEEE Trans Biomed Eng*, vol. 45, pp. 563–71, May 1998.

[177] M. Wright, E. Harks, S. Deladi, F. Suijver, M. Barley, A. van Dusschoten, S. Fokkenrood, F. Zuo, F. Sacher, M. Hocini, M. Haïssaguerre, and P. Jaïs, "Real-time lesion assessment using a novel combined ultrasound and radiofrequency ablation catheter," *Heart Rhythm*, vol. 8, pp. 304–12, Feb 2011.

[178] M. K. Jain and P. D. Wolf, "Temperature-controlled and constant-power radio-frequency ablation: what affects lesion growth?," *IEEE Trans Biomed Eng*, vol. 46, pp. 1405–12, Dec 1999.

[179] S. Tungjitkusolmun, V. R. Vorperian, N. Bhavaraju, H. Cao, J. Z. Tsai, and J. G. Webster, "Guidelines for predicting lesion size at common

endocardial locations during radio-frequency ablation," *IEEE Trans Biomed Eng*, vol. 48, pp. 194–201, Feb 2001.

[180] A. Price, Z. Leshen, J. Hansen, I. Singh, P. Arora, J. Koblish, and B. Avitall, "Novel ablation catheter technology that improves mapping resolution and monitoring of lesion maturation," *Journal of Innovations in Cardiac Rhythm Management*, vol. 3, pp. 599–609, 2012.

List of Publications and Supervised Thesis

Journal Articles

- **M. W. Keller**, S. Schuler, M. Wilhelms, G. Lenis, G. Seemann, C. Schmitt, O. Dössel, A. Luik, "Characterization of Radiofrequency Ablation Lesion Development Based on Simulated and Measured Intracardiac Electrograms" *IEEE Transactions on Biomedical Engineering*, 2014, *(published online ahead of print)*

Conference Contributions

- **M. W. Keller**, S. Schuler, A. Luik, G. Seemann, C. Schilling, C. Schmitt, and O. Dössel, "Comparison of simulated and clinical intracardiac electrograms," in *proceedings of the 35th Annual International Conference of the IEEE EMBS*, 2013, pp. 6858–6861.
- **M. W. Keller**, A. Luik, M. Abady, G. Seemann, C. Schmitt, and O. Dössel, "Influence of three-dimensional fibrotic patterns on simulated intracardiac electrogram morphology," in *Computing in Cardiology*, 2013, Vol. 40 pp. 923–926.
- **M. W. Keller**, S. Schuler, A. Luik, C. Schmitt, and O. Dössel, "Evaluating changes in electrogram morphology during radiofrequency ablation of cardiac arrhythmias," in *Biomedical Engineering/Biomedizinische Technik*, 2013, Vol. 58.

- **M. W. Keller**, S. Schuler, G. Seemann, and O. Dössel, "Differences in intracardiac signals on a realistic catheter geometry using mono and bidomain models," in *Computing in Cardiology*, 2012, Vol. 39, pp. 305–308.

- **M. W. Keller**, G. Seemann, and O. Dössel, "Simulating extracellular microelectrode recordings on cardiac tissue preparations in a bidomain model," in *Biomedizinische Technik / Biomedical Engineering*, 2012, Vol. 57

- **M. W. Keller**, and O. Dössel, "Towards simultaneous optical and electrical characterization of the electrode tissue interface in catheter measurements of atrial electrophysiology," in *Biomedizinische Technik / Biomedical Engineering*, 2011, Vol. 56.

- **M. W. Keller**, C. Schilling, A. Luik, C. Schmitt, and O. Dössel, "Descriptors for a classification of complex fractionated atrial electrograms as a guidance for catheter ablation of atrial fibrillation," in *Biomedizinische Technik / Biomedical Engineering*, 2010, Vol. 55, pp. 100–103.

- S. Schuler, **M. W. Keller**, T. Oesterlein, G. Seemann, and O. Dössel, "Influence of Catheter Orientation, Tissue Thickness and Conduction Velocity on the Intracardiac Electrogram," in *Biomedical Engineering/Biomedizinische Technik*, 2013, Vol. 58.

- T. G. Oesterlein, **M. W. Keller**, S. Schuler, A. Luik, G. Seemann, C. Schmitt, and O. Dössel, "Determination of local activation time in bipolar endocardial electrograms: a comparison of clinical criteria and a new method based on the non-linear energy operator," in *Journal of Electrocardiology*, 2013, Vol. 46, e18.

- C. Schilling, A. Luik, **M. W. Keller**, C. Schmitt, and O. Dössel, "Characterizing continuous activity with high fractionation during atrial fibrillation," in *Proc. 7th International Workshop on Biosignal Interpretation*, 2012, pp. 49–52.

Reports and Theses

- **M. Keller**, *"Charakterisierung und Analyse von räumlich-zeitlichen Erregungsmustern bei Vorhofflimmern,"* Diploma Thesis, Institute of Biomedical Engineering, Karlsruhe Institute of Technology (KIT), 2010.
- **M. Keller**, *"Merkmalsextraktion aus dem Signal eines Erste-Hilfe-Sensors mithilfe der Wigner-Ville-Verteilung,"* Student Research Project, Institute of Biomedical Engineering, Karlsruhe Institute of Technology (KIT), 2009.

Supervised Student Theses

- H. Gao, *"Aufbau und Ansteuerung einer High-Power-LED-Lichtquelle, für die fluoreszenzoptische Messung der Transmembranspannung von Kardiomyozyten,"* Bachelor's thesis, Institute of Biomedical Engineering, Karlsruhe Institute of Technology (KIT), Karlsruhe, 2012.
- F. Schäuble, *"Extraktion und Analyse von Aktivierungszeiten aus fluoreszenzoptischen Messdaten,"* Bachelor Thesis, Institute of Biomedical Engineering, Karlsruhe Institute of Technology (KIT), 2012.
- J. Lür Tischer, *"Entwicklung und Aufbau einer Messhardware zur Erfassung von extrazellulären Potenzialen an vitalem Myokardgewebe,"* Diploma Thesis, Institute of Biomedical Engineering, Karlsruhe Institute of Technology (KIT), 2013.
- S. Schuler, *"Simulation von intrakardialen Elektrogrammen während der Katheterablation,"* Bachelor Thesis, Institute of Biomedical Engineering, Karlsruhe Institute of Technology (KIT), 2012.
- M. Soltan Abady, *"Repräsentation von Gewebeveränderungen im intrakardialen Elektrogramm – eine Simulationsstudie,"* Master Thesis, Institute of Biomedical Engineering, Karlsruhe Institute of Technology (KIT), 2013.

Awards & Grants

- Further Education Scholarship ('Weiterqualifizierungsstipendium'), Karlsruhe House of Young Scientists (KHYS), 2012
- Networking Scholarship ('Kontaktstipendium'), Karlsruhe House of Young Scientists (KHYS), 2011

Karlsruhe Transactions on Biomedical Engineering
(ISSN 1864-5933)

Karlsruhe Institute of Technology / Institute of Biomedical Engineering (Ed.)

Die Bände sind unter www.ksp.kit.edu als PDF frei verfügbar oder als Druckausgabe bestellbar.

Band 2   Matthias Reumann
Computer assisted optimisation on non-pharmacological treatment of congestive heart failure and supraventricular arrhythmia. 2007
ISBN 978-3-86644-122-4

Band 3   Antoun Khawaja
Automatic ECG analysis using principal component analysis and wavelet transformation. 2007
ISBN 978-3-86644-132-3

Band 4   Dmytro Farina
Forward and inverse problems of electrocardiography: clinical investigations. 2008
ISBN 978-3-86644-219-1

Band 5   Jörn Thiele
Optische und mechanische Messungen von elektro-physiologischen Vorgängen im Myokardgewebe. 2008
ISBN 978-3-86644-240-5

Band 6   Raz Miri
Computer assisted optimization of cardiac resynchronization therapy. 2009
ISBN 978-3-86644-360-0

Band 7   Frank Kreuder
2D-3D-Registrierung mit Parameterentkopplung für die Patientenlagerung in der Strahlentherapie. 2009
ISBN 978-3-86644-376-1

Band 8   Daniel Unholtz
Optische Oberflächensignalmessung mit Mikrolinsen-Detektoren für die Kleintierbildgebung. 2009
ISBN 978-3-86644-423-2

Band 9   Yuan Jiang
Solving the inverse problem of electrocardiography in a realistic environment. 2010
ISBN 978-3-86644-486-7

Karlsruhe Transactions on Biomedical Engineering
(ISSN 1864-5933)

Band 10   Sebastian Seitz
           **Magnetic Resonance Imaging on Patients with Implanted
           Cardiac Pacemakers.** 2011
           ISBN 978-3-86644-610-6

Band 11   Tobias Voigt
           **Quantitative MR Imaging of the Electric Properties and Local
           SAR based on Improved RF Transmit Field Mapping.** 2011
           ISBN 978-3-86644-598-7

Band 12   Frank Michael Weber
           **Personalizing Simulations of the Human Atria: Intracardiac Measure-
           ments, Tissue Conductivities, and Cellular Electrophysiology.** 2011
           ISBN 978-3-86644-646-5

Band 13   David Urs Josef Keller
           **Multiscale Modeling of the Ventricles: from Cellular
           Electrophysio-logy to Body Surface Electrocardiograms.** 2011
           ISBN 978-3-86644-714-1

Band 14   Oussama Jarrousse
           **Modified Mass-Spring System for Physically Based
           Deformation Modeling.** 2012
           ISBN 978-3-86644-742-4

Band 15   Julia Bohnert
           **Effects of Time-Varying Magnetic Fields in the Frequency Range
           1 kHz to 100 kHz upon the Human Body: Numerical Studies and
           Stimulation Experiment.** 2012
           ISBN 978-3-86644-782-0

Band 16   Hanno Homann
           SAR Prediction and SAR Management for Parallel Transmit MRI. 2012
           ISBN 978-3-86644-800-1

Band 17   Christopher Schilling
           Analysis of Atrial Electrograms. 2012
           ISBN 978-3-86644-894-0

Band 18   Tobias Baas
           **ECG Based Analysis of the Ventricular Repolarisation
           in the Human Heart.** 2012
           ISBN 978-3-86644-882-7

Karlsruhe Transactions on Biomedical Engineering
(ISSN 1864-5933)

Band 19   Martin Wolfgang Krüger
**Personalized Multi-Scale Modeling of the Atria: Heterogeneities,
Fiber Architecture, Hemodialysis and Ablation Therapy.** 2012
ISBN 978-3-86644-948-0

Band 20   Mathias Wilhelms
**Multiscale Modeling of Cardiac Electrophysiology: Adaptation
to Atrial and Ventricular Rhythm Disorders and Pharmacological
Treatment.** 2013
ISBN 978-3-7315-0045-2

Band 21   Matthias Keller
**Formation of Intracardiac Electrograms under Physiological
and Pathological Conditions.** 2014
ISBN 978-3-7315-0228-9